Classical
Electromagnetism

Table of Contents

It is with great pleasure that I present this book. It has been carefully written after numerous discussions with my peers and other practitioners of the field. I would like to take this opportunity to thank my family and friends who have been extremely supporting at every step in my life.

The branch of theoretical physics which studies the interaction between currents and electric charges is known as classical electromagnetism. It primarily uses an extension of the classical Newtonian model. It is used to provide a description of electromagnetic phenomena at relatively large length scales and field strengths. Some of the fundamental concepts within this area of study are Liénard–Wiechert potentials and Jefimenko's equations. The classical electromagnetic effect of a moving electric point charge in terms of a scalar potential and vector potential in the Lorenz gauge is described through the Liénard–Wiechert potential. A few of the other elements of study within this field are electromagnetic waves, Lorentz force and the electric field. The topics included in this book on classical electromagnetism are of utmost significance and bound to provide incredible insights to readers. While understanding the long-term perspectives of the topics, it makes an effort in highlighting their impact as a modern tool for the growth of the discipline. This textbook is appropriate for students seeking detailed information in this area as well as for experts.

The chapters below are organized to facilitate a comprehensive understanding of the subject:

Chapter – What is Electromagnetism?

The branch of physics which is concerned with the study of the electromagnetic force is referred to as electromagnetism. The electromagnetic force is a type of physical interaction that takes place between electrically charged particles. This is an introductory chapter which will introduce briefly all the significant aspects of electromagnetism.

Chapter – Electrostatics

The branch of physics which deals with the study of electric charges at rest is known as electrostatics. Some of the laws studied within this field are Coulomb's law and Gauss's law. This chapter has been carefully written to provide an easy understanding of these fundamental concepts within electrostatics.

Chapter – Magnetostatics

The study of magnetic fields in the systems where the currents are steady is known as magnetostatics. Some of the important concepts within this field are Gauss's law for magnetism, Ampère's circuital law and magnetic dipole moment. This chapter closely examines these key concepts of magnetostatics.

Chapter – Electrodynamics

The study of the phenomena associated with charged bodies in motion, and fluctuating electric and magnetic fields is referred to as electrodynamics. There are various forces studied within this field such as Lorentz force, Electromotive force and Magnetomotive force. This chapter has been carefully written to provide an easy understanding of these aspects of electrodynamics.

Chapter – Electrical Circuits

The paths used for transmitting electric current are known as electrical circuits. Some of the theorems and laws studied in relation to electrical circuits are Ohm's law, Kirchhoff's circuit laws, Thévenin's theorem, Norton's theorem and superposition theorem. This chapter closely examines these major concepts associated with electric circuits.

Rebecca Williams

1
What is Electromagnetism?

The branch of physics which is concerned with the study of the electromagnetic force is referred to as electromagnetism. The electromagnetic force is a type of physical interaction that takes place between electrically charged particles. This is an introductory chapter which will introduce briefly all the significant aspects of electromagnetism.

Electromagnetism is the physics of the electromagnetic field: A field that exerts a force on particles that possess the property of electric charge, and it is in turn affected by the presence and motion of those particles.

A changing magnetic field produces an electric field, a phenomenon known as "electromagnetic induction." This phenomenon forms the basis of operation for electrical generators, induction motors, and transformers). Similarly, a changing electric field generates a magnetic field. A magnetic field is produced by the motion of electric charges, that is, an electric current. The magnetic field produces the magnetic force associated with magnets. Because of this interdependence of the electric and magnetic fields, it is appropriate to consider them as a single coherent entity, the electromagnetic field.

The theoretical implications of electromagnetism led to development of the theory of special relativity by Albert Einstein in 1905.

The Electromagnetic Force

The force that the electromagnetic field exerts on electrically charged particles, called the electromagnetic force, is one of the fundamental forces, and is responsible for most of the forces we experience in our daily lives. The other fundamental forces are the strong nuclear force (which holds atomic nuclei together), the weak nuclear force and the gravitational force. All other forces are ultimately derived from these fundamental forces.

The electromagnetic force is the one responsible for practically all the phenomena encountered in daily life, with the exception of gravity. All the forces involved in interactions between atoms can be traced to the electromagnetic force acting on the electrically charged protons and electrons inside the atoms. This includes the forces we experience in "pushing" or "pulling" ordinary material objects, which come from the intermolecular forces between the individual molecules in our bodies and those in the objects. It also includes all forms of chemical phenomena, which arise from interactions between electron orbitals.

Classical Electrodynamics

The scientist William Gilbert proposed, in his De Magnete (1600), that electricity and magnetism, while both capable of causing attraction and repulsion of objects, were distinct effects. Mariners had noticed that lightning strikes had the ability to disturb a compass needle, but the link between lightning and electricity was not confirmed until Benjamin Franklin's proposed experiments in 1752. One of the first to discover and publish a link between human-made electric current and magnetism was Romagnosi, who in 1802 noticed that connecting a wire across a Voltaic pile deflected a nearby compass needle. However, the effect did not become widely known until 1820, when Ørsted performed a similar experiment. Ørsted's work influenced Ampère to produce a theory of electromagnetism that set the subject on a mathematical foundation.

An accurate theory of electromagnetism, known as classical electromagnetism, was developed by various physicists over the course of the nineteenth century, culminating in the work of James Clerk Maxwell, who unified the preceding developments into a single theory and discovered the electromagnetic nature of light. In classical electromagnetism, the electromagnetic field obeys a set of equations known as Maxwell's equations, and the electromagnetic force is given by the Lorentz force law.

One of the peculiarities of classical electromagnetism is that it is difficult to reconcile with classical mechanics, but it is compatible with special relativity. According to Maxwell's equations, the speed of light in a vacuum is a universal constant, dependent only on the electrical permittivity and magnetic permeability of free space. This violates Galilean invariance, a long-standing cornerstone of classical mechanics. One way to reconcile the two theories is to assume the existence of a luminiferous aether through which the light propagates. However, subsequent experimental efforts failed to detect the presence of the aether. After important contributions of Hendrik Lorentz and Henri Poincaré, in 1905, Albert Einstein solved the problem with the introduction of special relativity, which replaces classical kinematics with a new theory of kinematics that is compatible with classical electromagnetism.

In addition, relativity theory shows that in moving frames of reference a magnetic field transforms to a field with a nonzero electric component and vice versa; thus firmly showing that they are two sides of the same coin, and thus the term "electromagnetism."

The Photoelectric Effect

In another paper published in that same year, Albert Einstein undermined the very foundations of classical electromagnetism. His theory of the photoelectric effect (for which he won the Nobel prize for physics) posited that light could exist in discrete particle-like quantities, which later came to be known as photons. Einstein's theory of the photoelectric effect extended the insights that appeared in the solution of the ultraviolet catastrophe presented by Max Planck in 1900. In his work, Planck showed that hot objects emit electromagnetic radiation in discrete packets, which leads to a finite total energy emitted as black body radiation. Both of these results were in direct contradiction with the classical view of light as a continuous wave. Planck's and Einstein's theories were progenitors of quantum mechanics, which, when formulated in 1925, necessitated the invention of a quantum theory of electromagnetism. This theory, completed in the 1940s, is known as quantum electrodynamics (or "QED"), and is one of the most accurate theories known to physics.

The term electrodynamics is sometimes used to refer to the combination of electromagnetism with mechanics, and deals with the effects of the electromagnetic field on the dynamic behavior of electrically charged particles.

Units

Electromagnetic units are part of a system of electrical units based primarily upon the magnetic properties of electric currents, the fundamental cgs unit being the ampere. The units are:

- Ampere (current)

- Coulomb (charge)

- Farad (capacitance)

- Henry (inductance)

- Ohm (resistance)

- Volt (electric potential)

- Watt (power)

In the electromagnetic cgs system, electrical current is a fundamental quantity defined via Ampère's law and takes the permeability as a dimensionless quantity (relative permeability) whose value in a vacuum is unity. As a consequence, the square of the speed of light appears explicitly in some of the equations interrelating quantities in this system.

SI Electromagnetism Units

Symbol	Name of Quantity	Derived Units	Unit	Base Units
I	Current	ampere (SI base unit)	A	$A = W/V = C/s$
q	Electric charge, Quantity of electricity	coulomb	C	$A \cdot s$
V	Potential difference	volt	V	$J/C = kg \cdot m^2 \cdot s^{-3} \cdot A^{-1}$
R, Z, X	Resistance, Impedance, Reactance	ohm	Ω	$V/A = kg \cdot m^2 \cdot s^{-3} \cdot A^{-2}$
ρ	Resistivity	ohm metre	$\Omega \cdot m$	$kg \cdot m^3 \cdot s^{-3} \cdot A^{-2}$
P	Power, Electrical	watt	W	$V \cdot A = kg \cdot m^2 \cdot s^{-3}$
C	Capacitance	farad	F	$C/V = kg^{-1} \cdot m^{-2} \cdot A^2 \cdot s^4$
	Elastance	reciprocal farad	F^{-1}	$V/C = kg \cdot m^2 \cdot A^{-2} \cdot s^{-4}$
ε	Permittivity	farad per metre	F/m	$kg^{-1} \cdot m^{-3} \cdot A^2 \cdot s^4$
χ_e	Electric susceptibility	(dimensionless)	-	-

G, Y, B	Conductance, Admittance, Susceptance	siemens	S	$\Omega^{-1} = kg^{-1}{\cdot}m^{-2}{\cdot}s^3{\cdot}A^2$
σ	Conductivity	siemens per metre	S/m	$kg^{-1}{\cdot}m^{-3}{\cdot}s^3{\cdot}A^2$
H	Auxiliary magnetic field, magnetic field intensity	ampere per metre	A/m	$A{\cdot}m^{-1}$
Φ_m	Magnetic flux	weber	Wb	$V{\cdot}s = kg{\cdot}m^2{\cdot}s^{-2}{\cdot}A^{-1}$
B	Magnetic field, magnetic flux density, magnetic induction, magnetic field strength	tesla	T	$Wb/m^2 = kg{\cdot}s^{-2}{\cdot}A^{-1}$
	Reluctance	ampere-turns per weber	A/Wb	$kg^{-1}{\cdot}m^{-2}{\cdot}s^2{\cdot}A^2$
L	Inductance	henry	H	$Wb/A = V{\cdot}s/A = kg{\cdot}m^2{\cdot}s^{-2}{\cdot}A^{-2}$
μ	Permeability	henry per metre	H/m	$kg{\cdot}m{\cdot}s^{-2}{\cdot}A^{-2}$
χ_m	Magnetic susceptibility	(dimensionless)	-	-

Electromagnetic Phenomena

In the theory, electromagnetism is the basis for optical phenomena, as discovered by James Clerk Maxwell while he studied electromagnetic waves. Light, being an electromagnetic wave, has properties that can be explained through the Maxwell's equations, such as reflection, refraction, diffraction, and interference. Relativity is born on the electromagnetic fields, as shown by Albert Einstein when he tried to make the electromagnetic theory compatible with Planck's radiation formula.

Explanation of Electromagnetism with an Example

Permanent Magnetic speakers commonly used in TV's and Radios are perfect example of Electromagnetic devices. Let's see the operation of these devices which are based on the principle of electromagnetism.

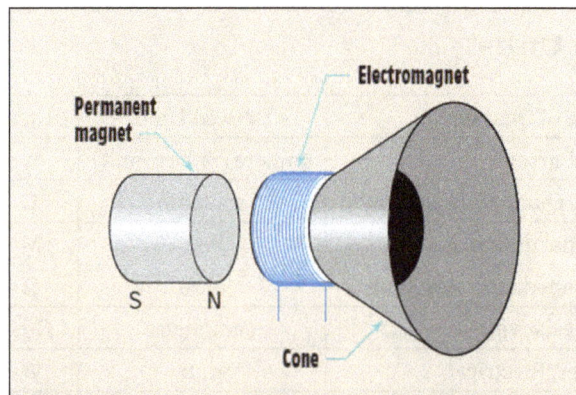

In order to convert electrical waves into audible sound, the speakers are designed. A metal coil is attached to a permanent magnet and when current passes through the coil it generates a magnetic field. The newly formed magnetic field is repelled by the permanent magnetic field resulting in the vibrations. These vibrations are amplified by the cone-like structure causing the sound. This is how speakers work based on electromagnetism.

Electromagnetic Induction

When a conductor is placed or moved through the magnetic field it generates voltage i.e., electricity. This principle is called Electromagnetic Induction. The voltages generated will be based on the speed of conductor moving through the electric field. Faster the speed of conductor, the greater the induced electricity or voltage.

Faraday's Law

Faraday's Law states that whenever there is relative motion between magnetic field and conductor, the flux linkage changes and this change in flux induces a voltage across the coil.

Explanation with an Example

DC Generator works on the principle of Faraday's Law of Electromagnetic Induction. It is a system that converts mechanical energy into electrical energy.

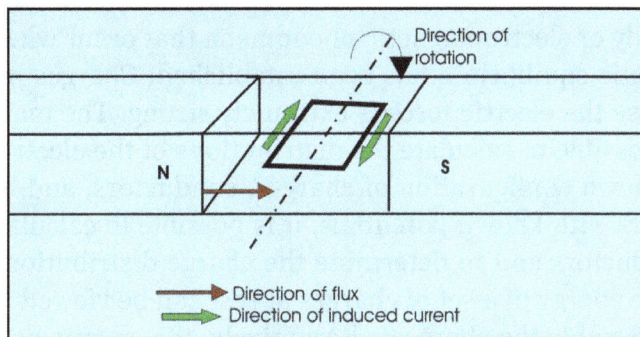

A rectangular conductor width sides are placed in between a magnetic field. When the rectangular conductor rotates in between magnetics, it cuts magnetic field thereby causing Electromagnetic field (e m f).

2
Electrostatics

The branch of physics which deals with the study of electric charges at rest is known as electrostatics. Some of the laws studied within this field are Coulomb's law and Gauss's law. This chapter has been carefully written to provide an easy understanding of these fundamental concepts within electrostatics.

Electrostatics is the study of electromagnetic phenomena that occur when there are no moving charges—i.e., after a static equilibrium has been established. Charges reach their equilibrium positions rapidly because the electric force is extremely strong. The mathematical methods of electrostatics make it possible to calculate the distributions of the electric field and of the electric potential from a known configuration of charges, conductors, and insulators. Conversely, given a set of conductors with known potentials, it is possible to calculate electric fields in regions between the conductors and to determine the charge distribution on the surface of the conductors. The electric energy of a set of charges at rest can be viewed from the standpoint of the work required to assemble the charges; alternatively, the energy also can be considered to reside in the electric field produced by this assembly of charges. Finally, energy can be stored in a capacitor; the energy required to charge such a device is stored in it as electrostatic energy of the electric field.

Electrical Conductor

Electrical conductor is a substance or material that allows electrons to flow atom to atom of that material with drift velocity in the conduction band against a small resistance offered by that substance.

Electrical Conductor may be metals, metal alloy, electrolyte or some non metals like graphite and conductive polymer.

Conduction of Current through a Conductor

The substance of the electrical conductor atom must have no energy gap between its valence band and conduction band. The outer electrons in the valence band are loosely attached to the atom.

When an electron gets excited due to electromotive force or thermal effect, it moves from its valence band to conduction band. Conduction band is the band where this electron gets its freedom to move anywhere in the conductor. The conductor is formed of atoms. Thus as a whole, the conduction band is in abundance of electrons.

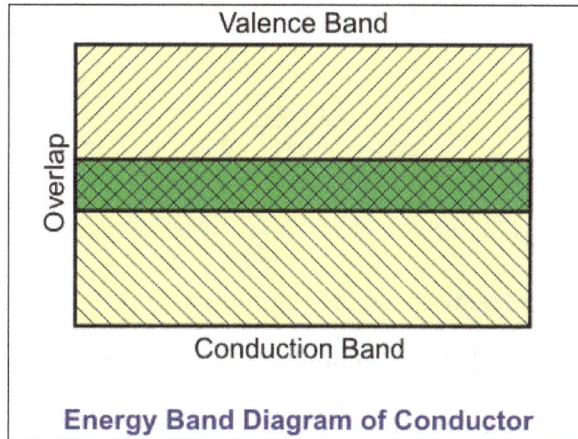

Energy Band Diagram of Conductor

In other word, it can be said that the metallic bonds are present in the conductors. These metallic bonds are based on structure of positive metal ions. These structures are surrounded by electron cloud.

When a potential difference occurs in the conductor across two points, the electrons get sufficient energy to flow from lower potency to higher potency in this conduction band against a small resistance offered by this conductor material. Electricity or current flows in the opposite direction of the flow of the electrons.

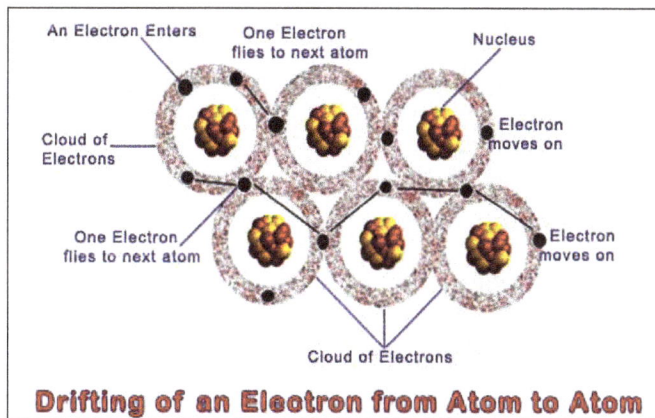

Drifting of an Electron from Atom to Atom

Electron Flow through a Conductor

Electrons do not move or flow in a straight line. In a conductor, the electrons are in to and fro motion or random velocity i.e. is called Drift Velocity (V_d) or average velocity. Due to this Drift Velocity, the electrons get collisions every moment with atoms or another electron in the conduction band of the conductor. Drift velocity is quite small, as there are so many free electrons. We can estimate the density of free electrons in a conductor, thus we can calculate the drift velocity for a given current. The larger the density, the lower the velocity required for a given current.

In the Conductor, flow of the electrons is against the Electric Field (E).

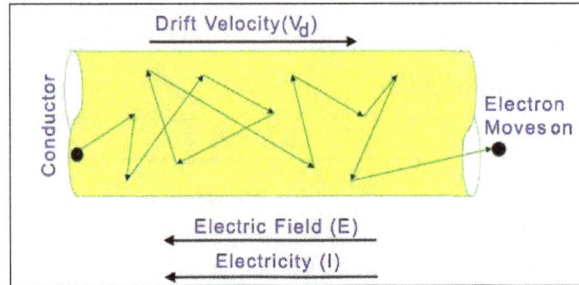

Properties of Conductor

The main properties that should be with a conductor are as follows:

1. A conductor always allows free movement of electrons or ions.

2. The electric field inside a conductor must be zero to permit the electrons or ions to move through the conductor.

3. Charge density inside a conductor is zero i.e. the positive and negative charges cancel inside a conductor.

4. As no charge inside the conductor, only free charges can exist only on the surface of a conductor.

5. The electric field is perpendicular to the surface of that conductor.

Type of the Conductors

Generally conductors can be classified based on Ohmic Response. They are:

Ohmic Conductors

This type of conductors always follow Ohm's Law ($V \propto I$)

V vs. I graph gives a straight line always.

Example: Aluminum, Silver, Copper etc.

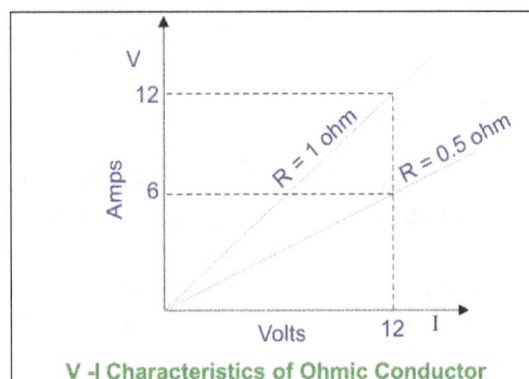

V -I Characteristics of Ohmic Conductor

Non Ohmic Conductors

This type of conductors never follow the Ohm's Law ($V \propto I$)

V vs. I graph does not give a straight line i.e. non linear graph.

Example: LDR (Light Dependant Resistor), Diode, Filament of Bulb, Thermistors etc.

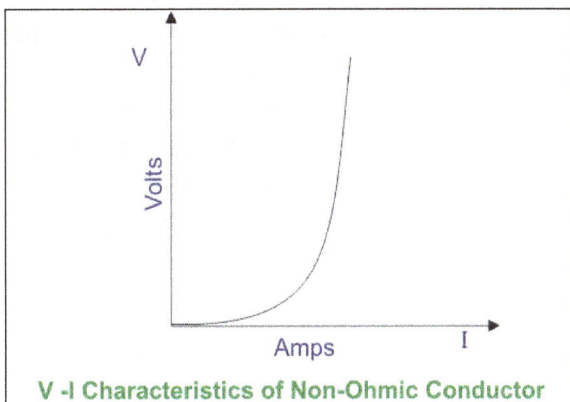

V -I Characteristics of Non-Ohmic Conductor

The examples of conductors are given below

Solid Conductor

1. Metallic Conductor: Silver, Copper, Aluminum, Gold etc.

2. Non Metallic Conductor: Graphite

3. Alloy Conductor: Brass, Bronze etc.

Liquid Conductor

1. Metallic Conductor: Mercury

2. Non Metallic Conductor: Saline Water, Acid Solution etc.

NB:

1. Copper Conductor is the most common material used for electrical wiring.

2. Gold Conductor is used for high-quality surface-to-surface contacts.

3. Silver is the best conductor in the Conductors list.

4. Impure Water is listed in Conductor List but it has less conductivity.

Charge of a Conductor Carrying Electricity

A current carrying conductor at any instance is with zero charge. It is because of at any instance number of electrons (at drift velocity) is equal to the number of protons in this conductor. So the net charge is zero.

Suppose a conductor is connected across a battery, i.e. positive end and negative end are connected with a conductor. Now electrons flow through the conductor from negative end to positive end of the battery. This flow of electrons is possible until this battery has EMF producing capability through chemical reaction inside.

Just think that here the conductor is the media through which charges can be passed from one electrode to another electrode of the battery. The electrons get rid of negative side of the battery and enter the conduction band of the conductor where already plenty of valence electrons of conductor atoms are available. The free electrons start journey in drift motion (towards positive electrode of the battery) from atom to atom in the conduction band. At any instance each atom holds zero charge because of drift electrons from adjacent atoms fill its valence band electron gaps and it happens continuously i.e. total number of electrons equal to the number of protons in the conductor at any moment. Now the rate of change of charge (q) with respect to time (t) is called current (I),

$$i.e.\ I = \frac{dq}{dt},$$

This rate of change of charge with respect to time occurs. Convention wise, Current (I) flows in opposite direction of electron flow.

When you remove the conductor from the battery, this conductor does not hold any charge particle, but EMF remains present across the battery electrodes with positive and negative polarity with no flow of electron.

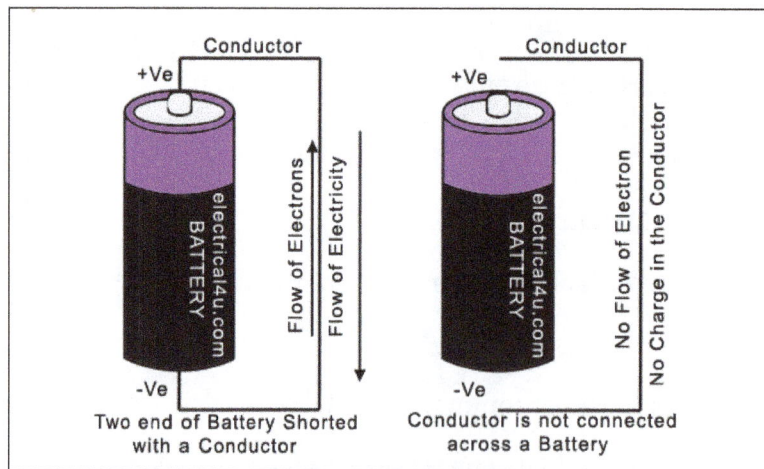

Effect of Temperature on a Conductor

The more effect of temperature, the more vibration in the conductor molecules. This impedes the electrons to flow, i.e. the electrons get obstruction to flow smoothly through the conductor. Thus, conductivity decreases gradually with increase the temperature.

Again, raise of temperature break some bonds in the conductor molecules and release some electrons. These electrons are less in number. As a whole, it can be said that increase in the temperature opposition against the drifting electron increases in the conductor.

Electrostatics of Conductors

Electrostatic Properties of Conductors

Conductors contain mobile charge carriers. In metallic conductors, these charge carriers are electrons. In a metal, the outer (valence) electrons part away from their atoms and are free to move. These electrons are free within the metal but not free to leave the metal.

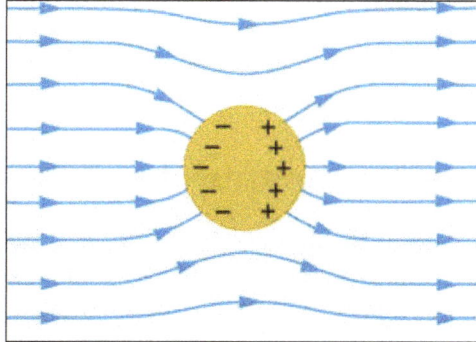

In an external electric field, they drift against the direction of the field. The positive ions made up of the nuclei and the bound electrons remain held in their fixed positions. In electrolytic conductors, the charge carriers are both positive and negative ions. Some of the important points about the electrostatic properties of a conductor are as follows:

The Electrostatic Field is Zero Inside a Conductor

In the static condition, whether a conductor is neutral or charged, the electric field inside the conductor is zero everywhere. This is one of the defining properties of a conductor.

We know that a conductor contains free electrons which, in the presence of an electric field, experience a drift or a force. Inside the conductor, the electrons distribute themselves in such a way that the final electric field at all points inside the conductor is zero.

Electrostatic Field Lines are Normal to the Surface at Every Point in a Charged Conductor

We can say, if the electric field lines were not normal at the surface, a component of the electric field would have been present along the surface of a conductor in the static condition.

Thus, free charges moving on the surface would also have experienced some force leading to their motion. But, this does not happen. Since there are no tangential components, the forces have to be normal to the surface.

In the Static Conditions, The Interior of the Conductor Contains No Excess Charge

We know that any neutral conductor contains an equal amount of positive and negative charges, at every point. This holds true even in an infinitesimally small element of volume or surface area.

From the Gauss's law, we can say that in the case of a charged conductor, the excess charges are present only on the surface.

Let us consider an arbitrary volume element of the conductor, which we denote as 'v' and for the closed surface bounding the volume element, the electrostatic field is zero. Thus, the total electric flux through S is zero. So, from the Gauss law, it follows that the net charge enclosed by the surface element is zero.

As we go on decreasing the size of the volume and the surface element, at a point we can say that when the element is vanishingly small, it denotes any point in the conductor. So the net charge at any point inside the conductor is always zero and the excess charges reside at the surface.

Constant Electrostatic Potential throughout the Volume of the Conductor

The electrostatic potential at any point throughout the volume of the conductor is always constant and the value of the electrostatic potential at the surface is equal to that at any point inside the volume.

Electric Charge

Electric charge is the physical property of matter that causes it to experience a force when placed in an electromagnetic field. There are two types of electric charge: *positive* and *negative* (commonly carried by protons and electrons respectively). Like charges repel and unlike attract. An object with an absence of net charge is referred to as *neutral*. Early knowledge of how charged substances interact is now called classical electrodynamics, and is still accurate for problems that do not require consideration of quantum effects.

Electric charge is a conserved property; the net charge of an isolated system, the amount of positive charge minus the amount of negative charge, cannot change. Electric charge is carried by subatomic particles. In ordinary matter, negative charge is carried by electrons, and positive charge is carried by the protons in the nuclei of atoms. If there are more electrons than protons in a piece of matter, it will have a negative charge, if there are fewer it will have a positive charge, and if there are equal numbers it will be neutral. Charge is *quantized*; it comes in integer multiples of individual small units called the elementary charge, e, about 1.602×10^{-19} coulombs, which is the smallest charge which can exist freely (particles called quarks have smaller charges, multiples of $1/3e$, but they are only found in combination, and always combine to form particles with integer charge). The proton has a charge of $+e$, and the electron has a charge of $-e$.

An electric charge has an electric field, and if the charge is moving it also generates a magnetic field. The combination of the electric and magnetic field is called the electromagnetic field, and its interaction with charges is the source of the electromagnetic force, which is one of the four fundamental forces in physics. The study of photon-mediated interactions among charged particles is called quantum electrodynamics.

The SI derived unit of electric charge is the coulomb (C) named after French physicist Charles-Augustin de Coulomb. In electrical engineering, it is also common to use the ampere-hour (Ah); in physics and chemistry, it is common to use the elementary charge (e as a unit). Chemistry also uses the Faraday constant as the charge on a mole of electrons. The symbol Q often denotes charge.

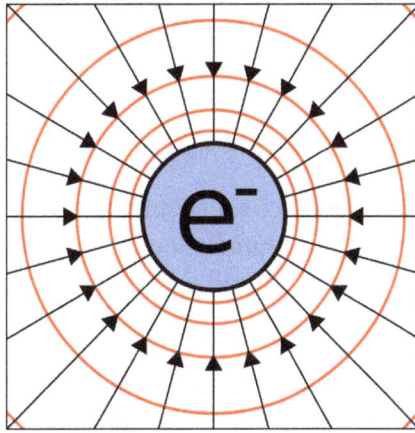

Diagram showing field lines and equipotentials around an electron, a negatively charged particle. In an electrically neutral atom, the number of electrons is equal to the number of protons (which are positively charged), resulting in a net zero overall charge.

Charge is the fundamental property of forms of matter that exhibit electrostatic attraction or repulsion in the presence of other matter. Electric charge is a characteristic property of many subatomic particles. The charges of free-standing particles are integer multiples of the elementary charge e; we say that electric charge is *quantized*. Michael Faraday, in his electrolysis experiments, was the first to note the discrete nature of electric charge. Robert Millikan's oil drop experiment demonstrated this fact directly, and measured the elementary charge. It has been discovered that one type of particle, quarks, have fractional charges of either $-\frac{1}{3}$ or $+\frac{2}{3}$, but it is believed they always occur in multiples of integral charge; free-standing quarks have never been observed.

By convention, the charge of an electron is negative, $-e$, while that of a proton is positive, $+e$. Charged particles whose charges have the same sign repel one another, and particles whose charges have different signs attract. Coulomb's law quantifies the electrostatic force between two particles by asserting that the force is proportional to the product of their charges, and inversely proportional to the square of the distance between them. The charge of an antiparticle equals that of the corresponding particle, but with opposite sign.

The electric charge of a macroscopic object is the sum of the electric charges of the particles that make it up. This charge is often small, because matter is made of atoms, and atoms typically have equal numbers of protons and electrons, in which case their charges cancel out, yielding a net charge of zero, thus making the atom neutral.

An *ion* is an atom (or group of atoms) that has lost one or more electrons, giving it a net positive charge (cation), or that has gained one or more electrons, giving it a net negative charge (anion). *Monatomic ions* are formed from single atoms, while *polyatomic ions* are formed from two or more atoms that have been bonded together, in each case yielding an ion with a positive or negative net charge.

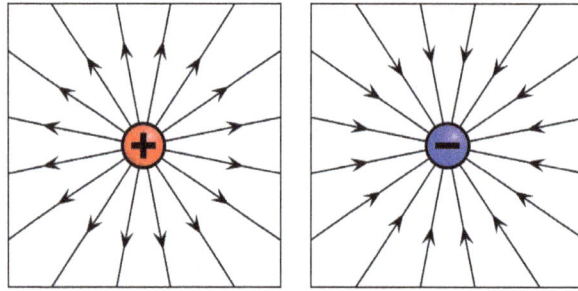

Electric field induced by a positive electric charge (left) and a field induced by a negative electric charge (right).

During formation of macroscopic objects, constituent atoms and ions usually combine to form structures composed of neutral *ionic compounds* electrically bound to neutral atoms. Thus macroscopic objects tend toward being neutral overall, but macroscopic objects are rarely perfectly net neutral.

Sometimes macroscopic objects contain ions distributed throughout the material, rigidly bound in place, giving an overall net positive or negative charge to the object. Also, macroscopic objects made of conductive elements, can more or less easily (depending on the element) take on or give off electrons, and then maintain a net negative or positive charge indefinitely. When the net electric charge of an object is non-zero and motionless, the phenomenon is known as static electricity. This can easily be produced by rubbing two dissimilar materials together, such as rubbing amber with fur or glass with silk. In this way non-conductive materials can be charged to a significant degree, either positively or negatively. Charge taken from one material is moved to the other material, leaving an opposite charge of the same magnitude behind. The law of *conservation of charge* always applies, giving the object from which a negative charge is taken a positive charge of the same magnitude, and vice versa.

Even when an object's net charge is zero, charge can be distributed non-uniformly in the object (e.g., due to an external electromagnetic field, or bound polar molecules). In such cases the object is said to be polarized. The charge due to polarization is known as bound charge, while charge on an object produced by electrons gained or lost from outside the object is called *free charge*. The motion of electrons in conductive metals in a specific direction is known as electric current.

Units

The SI derived unit of quantity of electric charge is the coulomb (symbol: C). The coulomb is defined as the quantity of charge that passes through the cross section of an electrical conductor carrying one ampere for one second. This unit was proposed in 1946 and ratified in 1948. In modern practice, the phrase "amount of charge" is used instead of "quantity of charge". The amount of charge in 1 electron (elementary charge) is approximately 1.6×10^{-19} C, and 1 coulomb corresponds to the amount of charge for about 6.24×10^{18} electrons. The symbol Q is often used to denote a quantity of electricity or charge. The quantity of electric charge can be directly measured with an electrometer, or indirectly measured with a ballistic galvanometer.

After finding the quantized character of charge, in 1891 George Stoney proposed the unit 'electron' for this fundamental unit of electrical charge. This was before the discovery of the particle by J. J.

Thomson in 1897. The unit is today treated as nameless, referred to as *elementary charge, fundamental unit of charge*, or simply as *e*. A measure of charge should be a multiple of the elementary charge *e*, even if at large scales charge seems to behave as a real quantity. In some contexts it is meaningful to speak of fractions of a charge; for example in the charging of a capacitor, or in the fractional quantum Hall effect.

The unit faraday is sometimes used in electrochemistry. One faraday of charge is the magnitude of the charge of one mole of electrons, i.e. 96485.33289(59) C.

In systems of units other than SI such as cgs, electric charge is expressed as combination of only three fundamental quantities (length, mass, and time), and not four, as in SI, where electric charge is a combination of length, mass, time, and electric current.

The Role of Charge in Static Electricity

Static electricity refers to the electric charge of an object and the related electrostatic discharge when two objects are brought together that are not at equilibrium. An electrostatic discharge creates a change in the charge of each of the two objects.

Electrification by Friction

When a piece of glass and a piece of resin—neither of which exhibit any electrical properties—are rubbed together and left with the rubbed surfaces in contact, they still exhibit no electrical properties. When separated, they attract each other.

A second piece of glass rubbed with a second piece of resin, then separated and suspended near the former pieces of glass and resin causes these phenomena:

- The two pieces of glass repel each other.

- Each piece of glass attracts each piece of resin.

- The two pieces of resin repel each other.

This attraction and repulsion is an *electrical phenomenon*, and the bodies that exhibit them are said to be *electrified*, or *electrically charged*. Bodies may be electrified in many other ways, as well as by friction. The electrical properties of the two pieces of glass are similar to each other but opposite to those of the two pieces of resin: The glass attracts what the resin repels and repels what the resin attracts.

If a body electrified in any manner whatsoever behaves as the glass does, that is, if it repels the glass and attracts the resin, the body is said to be *vitreously* electrified, and if it attracts the glass and repels the resin it is said to be *resinously* electrified. All electrified bodies are either vitreously or resinously electrified.

An established convention in the scientific community defines vitreous electrification as positive, and resinous electrification as negative. The exactly opposite properties of the two kinds of electrification justify our indicating them by opposite signs, but the application of the positive sign to one rather than to the other kind must be considered as a matter of arbitrary convention—just as it is a matter of convention in mathematical diagram to reckon positive distances towards the right hand.

No force, either of attraction or of repulsion, can be observed between an electrified body and a body not electrified.

The Role of Charge in Electric Current

Electric current is the flow of electric charge through an object, which produces no net loss or gain of electric charge. The most common charge carriers are the positively charged proton and the negatively charged electron. The movement of any of these charged particles constitutes an electric current. In many situations, it suffices to speak of the *conventional current* without regard to whether it is carried by positive charges moving in the direction of the conventional current or by negative charges moving in the opposite direction. This macroscopic viewpoint is an approximation that simplifies electromagnetic concepts and calculations.

At the opposite extreme, if one looks at the microscopic situation, one sees there are many ways of carrying an electric current, including: a flow of electrons; a flow of electron holes that act like positive particles; and both negative and positive particles (ions or other charged particles) flowing in opposite directions in an electrolytic solution or a plasma.

Beware that, in the common and important case of metallic wires, the direction of the conventional current is opposite to the drift velocity of the actual charge carriers; i.e., the electrons. This is a source of confusion for beginners.

Conservation of Electric Charge

The total electric charge of an isolated system remains constant regardless of changes within the system itself. This law is inherent to all processes known to physics and can be derived in a local form from gauge invariance of the wave function. The conservation of charge results in the charge-current continuity equation. More generally, the rate of change in charge density ρ within a volume of integration V is equal to the area integral over the current density J through the closed surface $S = \partial V$, which is in turn equal to the net current I:

$$-\frac{d}{dt} \int_V \rho \, dV = \oiint_{\partial V} J \cdot dS = \int J dS \, \cos\theta = I.$$

Thus, the conservation of electric charge, as expressed by the continuity equation, gives the result:

$$I = -\frac{dQ}{dt}.$$

The charge transferred between times t_i and t_f is obtained by integrating both sides:

$$Q = \int_{t_i}^{t_f} I \, dt$$

where I is the net outward current through a closed surface and Q is the electric charge contained within the volume defined by the surface.

Relativistic Invariance

Aside from the properties described in articles about electromagnetism, charge is a relativistic invariant. This means that any particle that has charge Q, no matter how fast it goes, always has charge Q. This property has been experimentally verified by showing that the charge of *one* helium nucleus (two protons and two neutrons bound together in a nucleus and moving around at high speeds) is the same as *two* deuterium nuclei (one proton and one neutron bound together, but moving much more slowly than they would if they were in a helium nucleus).

Coulomb's Law

Coulomb's law, or Coulomb's inverse-square law, is an experimental law of physics that quantifies the amount of force between two stationary, electrically charged particles. The electric force between charged bodies at rest is conventionally called *electrostatic force* or Coulomb force. The quantity of electrostatic force between stationary charges is always described by Coulomb's law. The law was first published in 1785 by French physicist Charles-Augustin de Coulomb, and was essential to the development of the theory of electromagnetism, maybe even its starting point, because it was now possible to discuss quantity of electric charge in a meaningful way. In its scalar form, the law is:

$$F = k_e \frac{q_1 q_2}{r^2},$$

where k_e is Coulomb's constant ($k_e \approx 9 \times 10^9$ N·m²·C⁻²), q_1 and q_2 are the signed magnitudes of the charges, and the scalar r is the distance between the charges. The force of the interaction between the charges is attractive if the charges have opposite signs (i.e., F is negative) and repulsive if like-signed (i.e., F is positive).

Being an inverse-square law, the law is analogous to Isaac Newton's inverse-square law of universal gravitation, but gravitational forces are always attractive, while electrostatic forces can be attractive or repulsive. Coulomb's law can be used to derive Gauss's law, and vice versa. In the case of a single stationary point charge, the two laws are equivalent, expressing the same physical law in different ways. The law has been tested extensively, and observations have upheld the law on a scale from 10^{-16} m to 10^8 m.

The Law

Coulomb's law states that: The magnitude of the electrostatic force of attraction or repulsion between two point charges is directly proportional to the product of the magnitudes of charges and inversely proportional to the square of the distance between them.

The force is along the straight line joining them. If the two charges have the same sign, the electrostatic force between them is repulsive; if they have different signs, the force between them is attractive.

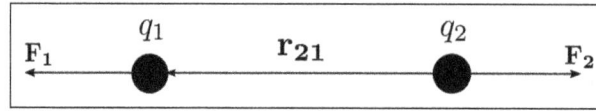

Coulomb's law can also be stated as a simple mathematical expression. The scalar and vector forms of the mathematical equation are:

$$|\mathrm{F}| = k_e \frac{|q_1 q_2|}{r^2} \qquad \text{and} \qquad F_1 = k_e \frac{q_1 q_2}{|r_{21}|^2} \hat{\mathbf{r}}_{21}, \qquad \text{respectively,}$$

where k_e is Coulomb's constant (k_e = $8.9875517873681764 \times 10^9$ N·m²·C⁻²), q_1 and q_2 are the signed magnitudes of the charges, the scalar r is the distance between the charges, the vector $\mathrm{r}_{21} = \mathrm{r}_1 - \mathrm{r}_2$ is the vectorial distance between the charges, and $\hat{\mathrm{r}}_{21} = \dfrac{r_{21}}{|r_{21}|}$ (a unit vector pointing from q_2 to q_1). The vector form of the equation calculates the force F_1 applied on q_1 by q_2. If r_{12} is used instead, then the effect on q_2 can be found. It can be also calculated using Newton's third law: $\mathrm{F}_2 = -\mathrm{F}_1$.

Units

When the electromagnetic theory is expressed in the International System of Units, force is measured in newtons, charge in coulombs, and distance in meters. Coulomb's constant is given by $k_e = \dfrac{1}{4\pi\varepsilon_0}$. The constant ε_0 is the vacuum electric permittivity (also known as "electric constant") in C²·m⁻²·N⁻¹. It should not be confused with ε_r, which is the dimensionless relative permittivity of the material in which the charges are immersed, or with their product $\varepsilon_a = \varepsilon_0 \varepsilon_r$, which is called "absolute permittivity of the material" and is still used in electrical engineering.

The SI derived units for the electric field are volts per meter, newtons per coulomb, or tesla meters per second.

Coulomb's law and Coulomb's constant can also be interpreted in various terms:

- Atomic units: In atomic units the force is expressed in hartrees per Bohr radius, the charge in terms of the elementary charge, and the distances in terms of the *Bohr radius*.

- Electrostatic units or Gaussian units: In electrostatic units and Gaussian units, the unit charge (*esu* or *statcoulomb*) is defined in such a way that the Coulomb constant k disappears because it has the value of one and becomes dimensionless.

- Lorentz–Heaviside units (also called *rationalized*): In Lorentz–Heaviside units the Coulomb constant is $k_e = 1/4\pi$ and becomes dimensionless.

Gaussian units and Lorentz–Heaviside units are both CGS unit systems. Gaussian units are more amenable for microscopic problems such as the electrodynamics of individual electrically charged particles. SI units are more convenient for practical, large-scale phenomena, such as engineering applications.

Electric Field

An electric field is a vector field that associates to each point in space the Coulomb force experienced by a test charge. In the simplest case, the field is considered to be generated solely by a single source point charge. The strength and direction of the Coulomb force F on a test charge q_t depends on the electric field E that it finds itself in, such that $F = q_t E$. If the field is generated by a positive source point charge q, the direction of the electric field points along lines directed radially outwards from it, i.e. in the direction that a positive point test charge q_t would move if placed in the field. For a negative point source charge, the direction is radially inwards.

The magnitude of the electric field E can be derived from Coulomb's law. By choosing one of the point charges to be the source, and the other to be the test charge, it follows from Coulomb's law that the magnitude of the electric field E created by a single source point charge q at a certain distance from it r in vacuum is given by:

$$|E| = \frac{1}{4\pi\varepsilon_0} \frac{|q|}{r^2}$$

Coulomb's Constant

Coulomb's constant is a proportionality factor that appears in Coulomb's law as well as in other electric-related formulas. The value of this constant is dependent upon the medium that the charged objects are immersed in. Denoted k_e, it is also called the electric force constant or electrostatic constant, hence the subscript e.

The exact value of Coulomb's constant in the case of air or vacuum is:

$$k_e = \frac{1}{4\pi\varepsilon_0} = \frac{c_0^2 \mu_0}{4\pi} = c_0^2 \times 10^{-7} \text{ H} \cdot \text{m}^{-1}$$
$$= 8.9875517873681764 \times 10^9 \text{ N} \cdot \text{m}^2 \cdot \text{C}^{-2}$$

Limitations

There are three conditions to be fulfilled for the validity of Coulomb's law:

- The charges must have a spherically symmetric distribution (e.g. be point charges, or a charged metal sphere).

- The charges must not overlap (e.g. they must be distinct point charges).

- The charges must be stationary with respect to each other.

The last of these is known as the electrostatic approximation. When movement takes place, Einstein's theory of relativity must be taken into consideration, and a result, an extra factor is introduced, which alters the force produced on the two objects. This extra part of the force is called the magnetic force, and is described by magnetic fields. For slow movement, the magnetic force is minimal and Coulomb's law can still be considered approximately correct, but when the charges are moving more quickly in relation to each other, the full electrodynamic rules (incorporating the magnetic force) must be considered.

Quantum Field Theory Origin

In simple terms, the Coulomb potential derives from the QED Lagrangian as follows. The Lagrangian of quantum electrodynamics is normally written in natural units, but in SI units, it is:

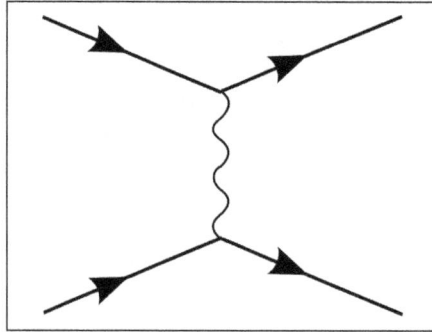

The most basic Feynman diagram for QED interaction between two fermions.

$$\mathcal{L}_{\text{QED}} = \bar{\psi}(i\hbar c\gamma^{\mu}D_{\mu} - mc^2)\psi - \frac{1}{4\hbar c}F_{\mu\nu}F^{\mu\nu}$$

where the covariant derivative (in SI units) is:

$$D_{\mu} = \partial_{\mu} + \frac{ig}{\hbar c}A_{\mu}$$

where g is the gauge coupling parameter. By putting the covariant derivative into the lagrangian explicitly, the interaction term (the term involving both A and ψ) is seen to be:

$$\mathcal{L}_{\text{int}} = ig\bar{\psi}\gamma^{\mu}A_{\mu}\psi$$

The most basic Feynman diagram for a QED interaction between two fermions is the exchange of a single photon, with no loops. Following the Feynman rules, this therefore contributes two QED vertex factors ($igQ\gamma_{\mu}$) to the potential, where Q is the QED-charge operator (Q gives the charge in terms of the electron charge, and hence is exactly –1 for electrons, etc.). For the photon in the diagram, the Feynman rules demand the contribution of one bosonic massless propagator $\left(\dfrac{\hbar c}{k^2}\right)$. Ignoring the momentum on the external legs (the fermions), the potential is therefore:

$$V(r) = \frac{1}{(2\pi)^3}\int e^{\frac{i\mathbf{k}\cdot\mathbf{r}}{\hbar}}(iQ_1 g^2\gamma_{\mu})(iQ_2 g^2\gamma_{\nu})\left(\frac{\hbar c}{k^2}\right)d^3k$$

which can be more usefully written as:

$$V(r) = \frac{-g^2\hbar c Q_1 Q_2}{4\pi}\frac{1}{(2\pi)^3}\int e^{\frac{i\mathbf{k}\cdot\mathbf{r}}{\hbar}}\frac{4\pi\eta_{\mu\nu}}{k^2}d^3k$$

where Q_i is the QED-charge on the ith particle. Recognising the integral as just being a Fourier transform enables the equation to be simplified:

$$V(r) = \frac{-g^2 \hbar c Q_1 Q_2}{4\pi} \frac{1}{r}.$$

For various reasons, it is more convenient to define the fine-structure constant $\alpha = \frac{g^2}{4\pi}$ and then define $e = \sqrt{4\pi\alpha\varepsilon_0 \hbar c}$ Rearranging these definitions gives:

$$g^2 \hbar c = \frac{e^2}{\varepsilon_0}$$

In natural units ($\hbar = 1, c = 1$ and $\varepsilon_0 = 1$), $g = e$. Continuing in SI units, the potential is therefore:

$$V(r) = \frac{-e^2}{4\pi\varepsilon_0} \frac{Q_1 Q_2}{r}$$

Defining $q_i = eQ_i$, as the macroscopic 'electric charge', makes 'e' the macroscopic 'electric charge' for an electron, and enables the formula to be put into the familiar form of the Coulomb potential:

$$V(r) = \frac{-1}{4\pi\varepsilon_0} \frac{q_1 q_2}{r}$$

The force $\left(\frac{dV(r)}{dr} \right)$ is therefore :

$$F(r) = \frac{1}{4\pi\varepsilon_0} \frac{q_1 q_2}{r^2}$$

The derivation makes clear that the force law is only an approximation — it ignores the momentum of the input and output fermion lines, and ignores all quantum corrections (i.e. the myriad possible diagrams with internal loops).

The Coulomb potential, and its derivation, can be seen as a special case of the Yukawa potential (specifically, the case where the exchanged boson – the photon – has no rest mass).

Scalar Form

When it is of interest to know the magnitude of the electrostatic force (and not its direction), it may be easiest to consider a scalar version of the law. The scalar form of Coulomb's Law relates the magnitude and sign of the electrostatic force F acting simultaneously on two point charges q_1 and q_2 as follows:

$$|F| = k_e \frac{|q_1 q_2|}{r^2}$$

where r is the separation distance and k_e is Coulomb's constant. If the product $q_1 q_2$ is positive, the force between the two charges is repulsive; if the product is negative, the force between them is attractive.

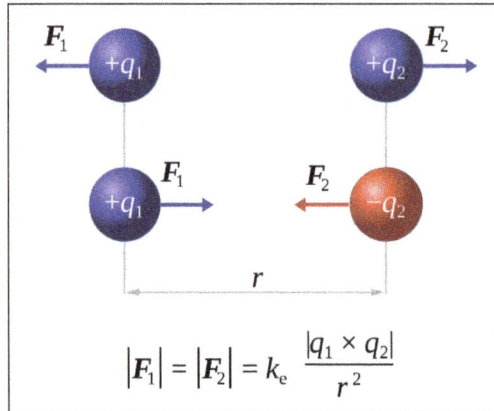

$$|F_1| = |F_2| = k_e \frac{|q_1 \times q_2|}{r^2}$$

The absolute value of the force F between two point charges q and Q relates to the distance between the point charges and to the simple product of their charges. The diagram shows that like charges repel each other, and opposite charges mutually attract.

Vector Form

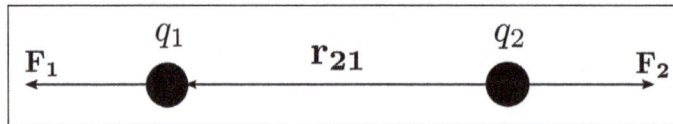

In the image, the vector F_1 is the force experienced by q_1, and the vector F_2 is the force experienced by q_2. When $q_1 q_2$ > 0 the forces are repulsive (as in the image) and when $q_1 q_2$ < 0 the forces are attractive (opposite to the image). The magnitude of the forces will always be equal.

Coulomb's law states that the electrostatic force F_1 experienced by a charge, q_1 at position r_1, in the vicinity of another charge, q_2 at position r_2, in a vacuum is equal to:

$$F_1 = \frac{q_1 q_2}{4\pi\varepsilon_0} \frac{(\mathbf{r_1} - \mathbf{r_2})}{|\mathbf{r_1} - \mathbf{r_2}|^3} = \frac{q_1 q_2}{4\pi\varepsilon_0} \frac{\hat{\mathbf{r}}_{21}}{|\mathbf{r_{21}}|^2},$$

where $r_{21} = r_1 - r_2$, the unit vector $\hat{r}_{21} = \dfrac{r_{21}}{|r_{21}|}$, and ε_0 is the electric constant.

The vector form of Coulomb's law is simply the scalar definition of the law with the direction given by the unit vector, $\hat{r}_{21}$, parallel with the line *from* charge q_2 to charge q_1. If both charges have the same sign (like charges) then the product $q_1 q_2$ is positive and the direction of the force on q_1 is given by $\hat{r}_{21}$; the charges repel each other. If the charges have opposite signs then the product $q_1 q_2$ is negative and the direction of the force on q_1 is given by $-\hat{r}_{21} = \hat{r}_{12}$; the charges attract each other.

The electrostatic force F_2 experienced by q_2, according to Newton's third law, is $F_2 = -F_1$.

System of Discrete Charges

The law of superposition allows Coulomb's law to be extended to include any number of point charges. The force acting on a point charge due to a system of point charges is simply the vector addition of the individual forces acting alone on that point charge due to each one of the charges. The resulting force vector is parallel to the electric field vector at that point, with that point charge removed.

The force F on a small charge q at position r, due to a system of N discrete charges in vacuum is:

$$\mathbf{F}(\mathbf{r}) = \frac{q}{4\pi\varepsilon_0} \sum_{i=1}^{N} q_i \frac{\mathbf{r}-\mathbf{r_i}}{|\mathbf{r}-\mathbf{r_i}|^3} = \frac{q}{4\pi\varepsilon_0} \sum_{i=1}^{N} q_i \frac{\widehat{\mathbf{R_i}}}{|\mathbf{R_i}|^2},$$

where q_i and r_i are the magnitude and position respectively of the ith charge, $\hat{R}_i$ is a unit vector in the direction of $R_i = r - r_i$ (a vector pointing from charges q_i to q).

Continuous Charge Distribution

In this case, the principle of linear superposition is also used. For a continuous charge distribution, an integral over the region containing the charge is equivalent to an infinite summation, treating each infinitesimal element of space as a point charge dq. The distribution of charge is usually linear, surface or volumetric.

For a linear charge distribution (a good approximation for charge in a wire) where $\lambda(r')$ gives the charge per unit length at position r', and $d\ell'$ is an infinitesimal element of length,

$$dq = \lambda(\mathbf{r}')\,d\ell'.$$

For a surface charge distribution (a good approximation for charge on a plate in a parallel plate capacitor) where $\sigma(r')$ gives the charge per unit area at position r', and dA' is an infinitesimal element of area,

$$dq = \sigma(\mathbf{r}')\,dA'.$$

For a volume charge distribution (such as charge within a bulk metal) where $\rho(r')$ gives the charge per unit volume at position r', and dV' is an infinitesimal element of volume,

$$dq = \rho(\mathbf{r}')\,dV'.$$

The force on a small test charge q' at position r in vacuum is given by the integral over the distribution of charge:

$$F = \frac{q'}{4\pi\varepsilon_0} \int dq \frac{\mathbf{r}-\mathbf{r}'}{|\mathbf{r}-\mathbf{r}'|^3}.$$

Simple Experiment to Verify Coulomb's Law

It is possible to verify Coulomb's law with a simple experiment. Consider two small spheres of mass m and same-sign charge q, hanging from two ropes of negligible mass of length l. The forces acting on each sphere are three: the weight mg, the rope tension T and the electric force F.

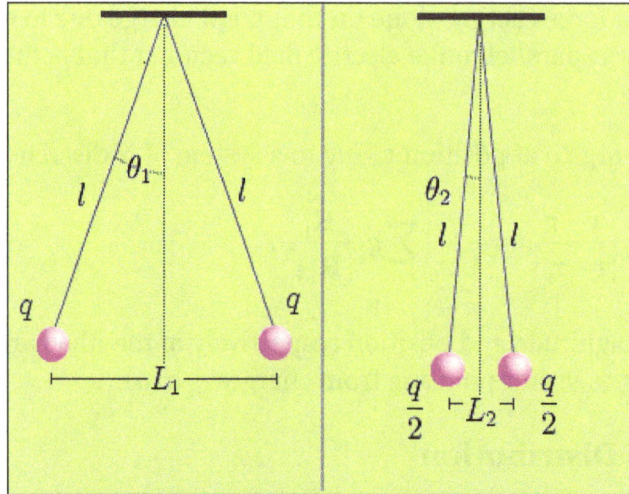

Experiment to verify Coulomb's law.

In the equilibrium state:

$$T \sin \theta_1 = F_1$$

and

$$T \cos \theta_1 = mg$$

Dividing $T \sin \theta_1 = F_1$ by $T \cos \theta_1 = mg$:

$$\frac{\sin \theta_1}{\cos \theta_1} = \frac{F_1}{mg} \Rightarrow F_1 = mg \tan \theta_1$$

Let L_1 be the distance between the charged spheres; the repulsion force between them F_1, assuming Coulomb's law is correct, is equal to,

$$F_1 = \frac{q^2}{4\pi\varepsilon_0 L_1^2}$$

so:

$$\frac{q^2}{4\pi\varepsilon_0 L_1^2} = mg \tan \theta_1$$

If we now discharge one of the spheres, and we put it in contact with the charged sphere, each one of them acquires a charge $\frac{q}{2}$. In the equilibrium state, the distance between the charges will be $L_2 < L_1$ and the repulsion force between them will be:

$$F_2 = \frac{(\frac{q}{2})^2}{4\pi\varepsilon_0 L_2^2} = \frac{\frac{q^2}{4}}{4\pi\varepsilon_0 L_2^2}$$

We know that $F_2 = mg \tan \theta_2$. And:

$$\frac{\frac{q^2}{4}}{4\pi\varepsilon_0 L_2^2} = mg \tan \theta_2$$

Dividing $\dfrac{q^2}{4\pi\varepsilon_0 L_1^2} = mg \tan \theta_1$ by $F_2 = \dfrac{(\frac{q}{2})^2}{4\pi\varepsilon_0 L_2^2} = \dfrac{\frac{q^2}{4}}{4\pi\varepsilon_0 L_2^2}$, we get:

$$\frac{\left(\dfrac{q^2}{4\pi\varepsilon_0 L_1^2}\right)}{\left(\dfrac{\frac{q^2}{4}}{4\pi\varepsilon_0 L_2^2}\right)} = \frac{mg \tan \theta_1}{mg \tan \theta_2} \Rightarrow 4\left(\frac{L_2}{L_1}\right)^2 = \frac{\tan \theta_1}{\tan \theta_2}$$

Measuring the angles θ_1 and θ_2 and the distance between the charges L_1 and L_2 is sufficient to verify that the equality is true taking into account the experimental error. In practice, angles can be difficult to measure, so if the length of the ropes is sufficiently great, the angles will be small enough to make the following approximation:

$$\tan \theta \approx \sin \theta = \frac{\frac{L}{2}}{\ell} = \frac{L}{2\ell} \Rightarrow \frac{\tan \theta_1}{\tan \theta_2} \approx \frac{\frac{L_1}{2\ell}}{\frac{L_2}{2\ell}}$$

Using this approximation, the relationship $4\left(\dfrac{L_2}{L_1}\right)^2 = \dfrac{\tan \theta_1}{\tan \theta_2}$ becomes the much simpler expression:

$$\frac{\frac{L_1}{2\ell}}{\frac{L_2}{2\ell}} \approx 4\left(\frac{L_2}{L_1}\right)^2 \Rightarrow \frac{L_1}{L_2} \approx 4\left(\frac{L_2}{L_1}\right)^2 \Rightarrow \frac{L_1}{L_2} \approx \sqrt[3]{4}$$

In this way, the verification is limited to measuring the distance between the charges and check that the division approximates the theoretical value.

Atomic Forces

Coulomb's law holds even within atoms, correctly describing the force between the positively charged atomic nucleus and each of the negatively charged electrons. This simple law also correctly accounts for the forces that bind atoms together to form molecules and for the forces that bind atoms and molecules together to form solids and liquids. Generally, as the distance between ions increases, the force of attraction, and binding energy, approach zero and ionic bonding is less favorable. As the magnitude of opposing charges increases, energy increases and ionic bonding is more favorable.

Electric Dipole Moment

The electric dipole moment is a measure of the separation of positive and negative electrical charges within a system, that is, a measure of the system's overall polarity. The SI units for electric dipole moment are coulomb-meter (C·m); however, a commonly used unit in atomic physics and chemistry is the debye (D).

Theoretically, an electric dipole is defined by the first-order term of the multipole expansion; it consists of two equal and opposite charges that are infinitesimally close together. This is unrealistic, as real dipoles have separated charge. However, because the charge separation is very small compared to everyday lengths, the error introduced by treating real dipoles like they are theoretically perfect is usually negligible. The dipole's direction usually points from the negative charge towards the positive charge. Direction is independent of the sides.

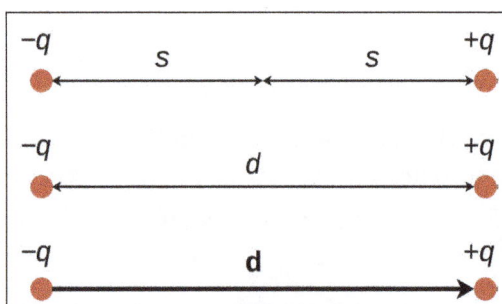

Quantities defining the electric dipole moment of two point charges.

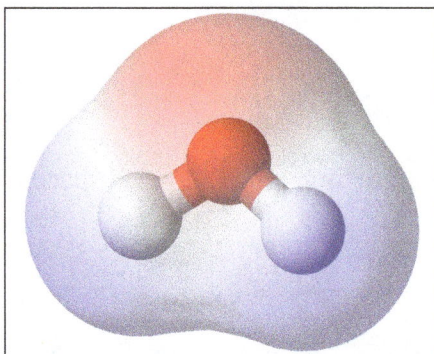

A molecule of water is polar because of the unequal sharing of its electrons in a "bent" structure. A separation of charge is present with negative charge in the middle (red shade), and positive charge at the ends (blue shade).

Often in physics the dimensions of a massive object can be ignored and can be treated as a point-like object, i.e. a point particle. Point particles with electric charge are referred to as point charges. Two point charges, one with charge $+q$ and the other one with charge $-q$ separated by a distance d, constitute an *electric dipole* (a simple case of an electric multipole). For this case, the electric dipole moment has a magnitude,

$$p = qd$$

and is directed from the negative charge to the positive one. Some may split d in half and use s = d/2 since this quantity is the distance between either charge and the center of the dipole, leading to a factor of two in the definition.

A stronger mathematical definition is to use vector algebra, since a quantity with magnitude and direction, like the dipole moment of two point charges, can be expressed in vector form

$$p = qd$$

where d is the displacement vector pointing from the negative charge to the positive charge. The electric dipole moment vector p also points from the negative charge to the positive charge.

An idealization of this two-charge system is the electrical point dipole consisting of two (infinite) charges only infinitesimally separated, but with a finite p.

This quantity is used in the definition of polarization density.

Energy and Torque

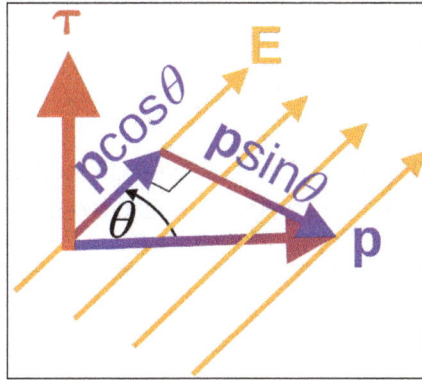

Electric dipole **p** and its torque **τ** in a uniform **E** field.

An object with an electric dipole moment is subject to a torque τ when placed in an external electric field. The torque tends to align the dipole with the field. A dipole aligned parallel to an electric field has lower potential energy than a dipole making some angle with it. For a spatially uniform electric field E, the energy U and the torque τ are given by:

$$U = -p \cdot E, \qquad \tau = p \times E,$$

where p is the dipole moment, and the symbol "×" refers to the vector cross product. The field vector and the dipole vector define a plane, and the torque is directed normal to that plane with the direction given by the right-hand rule.

A dipole oriented co- or anti-parallel to the direction in which a non-uniform electric field is increasing (gradient of the field) will experience a torque, as well as a force in the direction of its dipole moment. It can be shown that this force will always be parallel to the dipole moment regardless of co- or anti-parallel orientation of the dipole.

Expression

More generally, for a continuous distribution of charge confined to a volume V, the corresponding expression for the dipole moment is:

$$\mathbf{p}(\mathbf{r}) = \int_V \rho(\mathbf{r}_0)(\mathbf{r}_0 - \mathbf{r}) \, d^3\mathbf{r}_0,$$

where r locates the point of observation and d^3r_0 denotes an elementary volume in V. For an array of point charges, the charge density becomes a sum of Dirac delta functions:

$$\rho(\mathbf{r}) = \sum_{i=1}^{N} q_i \, \delta(\mathbf{r} - \mathbf{r}_i),$$

where each $\mathbf{r}_i$ is a vector from some reference point to the charge q_i. Substitution into the above integration formula provides:

$$\mathbf{p}(\mathbf{r}) = \sum_{i=1}^{N} q_i \int_{V} \delta(\mathbf{r}_0 - \mathbf{r}_i)(\mathbf{r}_0 - \mathbf{r}) \, d^3\mathbf{r}_0 = \sum_{i=1}^{N} q_i(\mathbf{r}_i - \mathbf{r}).$$

This expression is equivalent to the previous expression in the case of charge neutrality and $N = 2$. For two opposite charges, denoting the location of the positive charge of the pair as $\mathbf{r}_+$ and the location of the negative charge as r:

$$\mathbf{p}(\mathbf{r}) = q_1(\mathbf{r}_1 - \mathbf{r}) + q_2(\mathbf{r}_2 - \mathbf{r}) = q(\mathbf{r}_+ - \mathbf{r}) - q(\mathbf{r}_- - \mathbf{r}) = q(\mathbf{r}_+ - \mathbf{r}_-) = q\mathbf{d},$$

showing that the dipole moment vector is directed from the negative charge to the positive charge because the position vector of a point is directed outward from the origin to that point.

The dipole moment is particularly useful in the context of an overall neutral system of charges, for example a pair of opposite charges, or a neutral conductor in a uniform electric field. For such a system of charges, visualized as an array of paired opposite charges, the relation for electric dipole moment is:

$$p(r) = \sum_{i=1}^{N} \int_{V} q_i \left[\delta(r_0 - (r_i + d_i)) - \delta(r_0 - r_i) \right](r_0 - r) \, d^3 r_0$$

$$= \sum_{i=1}^{N} q_i \left[r_i + d_i - r - (r_i - r) \right]$$

$$= \sum_{i=1}^{N} q_i d_i = \sum_{i=1}^{N} p_i \, ,$$

where r is the point of observation, and $d_i = r_i - r'_i$, r_i being the position of the positive charge in the dipole i, and r'_i the position of the negative charge. This is the vector sum of the individual dipole moments of the neutral charge pairs. (Because of overall charge neutrality, the dipole moment is independent of the observer's position r.) Thus, the value of p is independent of the choice of reference point, provided the overall charge of the system is zero.

When discussing the dipole moment of a non-neutral system, such as the dipole moment of the proton, a dependence on the choice of reference point arises. In such cases it is conventional to choose the reference point to be the center of mass of the system, not some arbitrary origin. This choice is not only a matter of convention: the notion of dipole moment is essentially derived from the mechanical notion of torque, and as in mechanics, it is computationally and theoretically useful to choose the center of mass as the observation point. For a charged molecule the center of charge should be the reference point instead of the center of mass. For neutral systems the references point is not important. The dipole moment is an intrinsic property of the system.

Potential and Field of an Electric Dipole

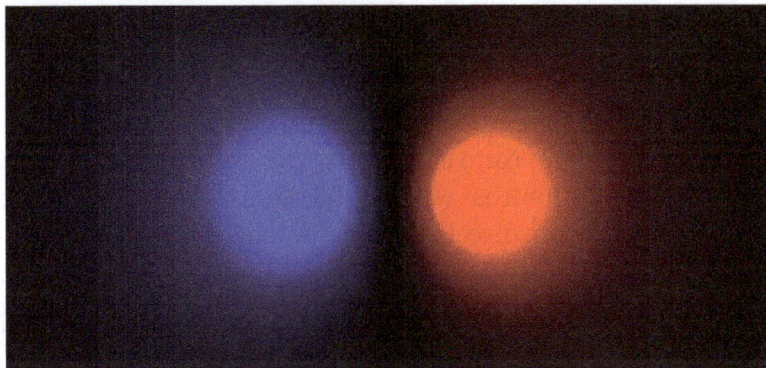

An electric dipole potential map. Negative potentials are in blue; positive potentials, in red.

An ideal dipole consists of two opposite charges with infinitesimal separation. The potential and field of such an ideal dipole are found next as a limiting case of an example of two opposite charges at non-zero separation.

Two closely spaced opposite charges have a potential of the form:

$$\phi(\mathbf{r}) = \frac{q}{4\pi\varepsilon_0 \left|\mathbf{r}-\mathbf{r}_+\right|} - \frac{q}{4\pi\varepsilon_0 \left|\mathbf{r}-\mathbf{r}_-\right|},$$

with charge separation, d, defined as,

$$d = r_+ - r_-,$$

The position relative to their center of mass (assuming equal masses), R, and the unit vector in the direction of R are given by:

$$R = r - \frac{\mathbf{r}_+ + \mathbf{r}_-}{2}, \quad \hat{R} = \frac{R}{R},$$

Taylor expansion in d/R allows this potential to be expressed as a series.

$$\phi(R) = \frac{1}{4\pi\varepsilon_0} \frac{qd \cdot \hat{R}}{R^2} + O\!\left(\frac{d^2}{R^2}\right) \approx \frac{1}{4\pi\varepsilon_0} \frac{p \cdot \hat{R}}{R^2},$$

where higher order terms in the series are vanishing at large distances, R, compared to d. Here, the electric dipole moment p is, as above:

$$p = qd .$$

The result for the dipole potential also can be expressed as:

$$\phi(R) = -p \cdot \nabla \frac{1}{4\pi\varepsilon_0 R},$$

which relates the dipole potential to that of a point charge. A key point is that the potential of the dipole falls off faster with distance R than that of the point charge.

The electric field of the dipole is the negative gradient of the potential, leading to:

$$E(R) = \frac{3(p \cdot \hat{R})\hat{R} - p}{4\pi\varepsilon_0 R^3}.$$

Thus, although two closely spaced opposite charges are not quite an ideal electric dipole (because their potential at short distances is not that of a dipole), at distances much larger than their separation, their dipole moment p appears directly in their potential and field.

As the two charges are brought closer together (d is made smaller), the dipole term in the multipole expansion based on the ratio d/R becomes the only significant term at ever closer distances R, and in the limit of infinitesimal separation the dipole term in this expansion is all that matters. As d is made infinitesimal, however, the dipole charge must be made to increase to hold p constant. This limiting process results in a "point dipole".

Dipole Moment Density and Polarization Density

The dipole moment of an array of charges,

$$p = \sum_{i=1}^{N} q_i \mathbf{d_i},$$

determines the degree of polarity of the array, but for a neutral array it is simply a vector property of the array with no information about the array's absolute location. The dipole moment *density* of the array p(r) contains both the location of the array and its dipole moment. When it comes time to calculate the electric field in some region containing the array, Maxwell's equations are solved, and the information about the charge array is contained in the *polarization density* P(r) of Maxwell's equations. Depending upon how fine-grained an assessment of the electric field is required, more or less information about the charge array will have to be expressed by P(r). sometimes it is sufficiently accurate to take P(r) = p(r). Sometimes a more detailed description is needed (for example, supplementing the dipole moment density with an additional quadrupole density) and sometimes even more elaborate versions of P(r) are necessary.

It now is explored just in what way the polarization density P(r) that enters Maxwell's equations is related to the dipole moment p of an overall neutral array of charges, and also to the *dipole moment density* p(r) (which describes not only the dipole moment, but also the array location). Only static situations are considered in what follows, so P(r) has no time dependence, and there is no displacement current.

A formulation of Maxwell's equations based upon division of charges and currents into "free" and "bound" charges and currents leads to introduction of the D- and P-fields:

$$D = \varepsilon_0 E + P,$$

where P is called the polarization density. In this formulation, the divergence of this equation yields:

$$\nabla \cdot D = \rho_f = \varepsilon_0 \nabla \cdot E + \nabla \cdot P,$$

and as the divergence term in E is the *total* charge, and ρ_f is "free charge", we are left with the relation:

$$\nabla \cdot P = -\rho_b,$$

with ρ_b as the bound charge, by which is meant the difference between the total and the free charge densities.

As an aside, in the absence of magnetic effects, Maxwell's equations specify that,

$$\nabla \times E = 0,$$

which implies,

$$\nabla \times (D - P) = 0,$$

Applying Helmholtz decomposition:

$$D - P = -\nabla \varphi,$$

for some scalar potential φ, and:

$$\nabla \cdot (D - P) = \varepsilon_0 \nabla \cdot E = \rho_f + \rho_b = -\nabla^2 \varphi.$$

Suppose the charges are divided into free and bound, and the potential is divided into,

$$\varphi = \varphi_f + \varphi_b.$$

Satisfaction of the boundary conditions upon φ may be divided arbitrarily between φ_f and φ_b because only the sum φ must satisfy these conditions. It follows that P is simply proportional to the electric field due to the charges selected as bound, with boundary conditions that prove convenient. In particular, when *no* free charge is present, one possible choice is P = ε_o E.

Medium with Charge and Dipole Densities

As described next, a model for polarization moment density p(r) results in a polarization,

$$P(r) = p(r)$$

restricted to the same model. For a smoothly varying dipole moment distribution p(r), the corresponding bound charge density is simply:

$$\nabla \cdot p(r) = \rho_b,$$

as we will establish shortly via integration by parts. However, if p(r) exhibits an abrupt step in dipole moment at a boundary between two regions, $\nabla \cdot$ p(r) results in a surface charge component of bound charge. This surface charge can be treated through a surface integral, or by using discontinuity conditions at the boundary.

As a first example relating dipole moment to polarization, consider a medium made up of a continuous charge density $\rho(r)$ and a continuous dipole moment distribution p(r). The potential at a position r is:

$$\phi(\mathbf{r}) = \frac{1}{4\pi\varepsilon_0} \int \frac{\rho(r_0)}{|r - r_0|} d^3 r_0 + \frac{1}{4\pi\varepsilon_0} \int \frac{p(r_0)\cdot(r - r_0)}{|r - r_0|^3} d^3 r_0,$$

where $\rho(r)$ is the unpaired charge density, and p(r) is the dipole moment density. Using an identity:

$$\nabla_{r_0} \frac{1}{|r - r_0|} = \frac{r - r_0}{|r - r_0|^3}$$

the polarization integral can be transformed:

$$\frac{1}{4\pi\varepsilon_0} \int \frac{p(r_0)\cdot(r - r_0)}{|r - r_0|^3} d^3 r_0 = \frac{1}{4\pi\varepsilon_0} \int p(r_0) \cdot \nabla_{r_0} \frac{1}{|r - r_0|} d^3 r_0,$$

$$= \frac{1}{4\pi\varepsilon_0} \int \nabla_{r_0} \cdot \left(p(r_0) \frac{1}{|r - r_0|} \right) d^3 r_0 - \frac{1}{4\pi\varepsilon_0} \int \frac{\nabla_{r_0} \cdot p(r_0)}{|r - r_0|} d^3 r_0,$$

The first term can be transformed to an integral over the surface bounding the volume of integration, and contributes a surface charge density, Putting this result back into the potential, and ignoring the surface charge for now:

$$\phi(r) = \frac{1}{4\pi\varepsilon_0} \int \frac{\rho(r_0) - \nabla_{r_0} \cdot p(r_0)}{|r - r_0|} d^3 r_0 ,$$

where the volume integration extends only up to the bounding surface, and does not include this surface.

The potential is determined by the total charge, which the above shows consists of:

$$\rho_{\text{total}}(r_0) = \rho(r_0) - \nabla_{r_0} \cdot p(r_0),$$

showing that:

$$-\nabla_{r_0} \cdot p(r_0) = \rho_b .$$

In short, the dipole moment density p(r) plays the role of the polarization density P for this medium. Notice, p(r) has a non-zero divergence equal to the bound charge density (as modeled in this approximation).

It may be noted that this approach can be extended to include all the multipoles: dipole, quadrupole, etc. Using the relation:

$$\nabla \cdot D = \rho_f \,,$$

the polarization density is found to be:

$$P(r) = p_{\text{dip}} - \nabla \cdot p_{\text{quad}} + \ldots,$$

where the added terms are meant to indicate contributions from higher multipoles. Evidently, inclusion of higher multipoles signifies that the polarization density P no longer is determined by a dipole moment density p alone. For example, in considering scattering from a charge array, different multipoles scatter an electromagnetic wave differently and independently, requiring a representation of the charges that goes beyond the dipole approximation.

Surface Charge

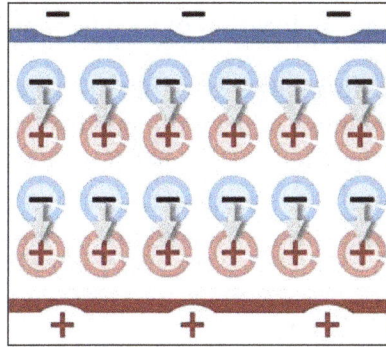

A uniform array of identical dipoles is equivalent to a surface charge.

Above, discussion was deferred for the first term in the expression for the potential due to the dipoles. Integrating the divergence results in a surface charge. The figure at the right provides an intuitive idea of why a surface charge arises. The figure shows a uniform array of identical dipoles between two surfaces. Internally, the heads and tails of dipoles are adjacent and cancel. At the bounding surfaces, however, no cancellation occurs. Instead, on one surface the dipole heads create a positive surface charge, while at the opposite surface the dipole tails create a negative surface charge. These two opposite surface charges create a net electric field in a direction opposite to the direction of the dipoles.

This idea is given mathematical form using the potential expression above. The potential is:

$$\phi(r) = \frac{1}{4\pi\varepsilon_0} \int \nabla_{r_0} \cdot \left(p(r_0) \frac{1}{|r - r_0|} \right) d^3 r_0 - \frac{1}{4\pi\varepsilon_0} \int \frac{\nabla_{r_0} \cdot p(r_0)}{|r - r_0|} d^3 r_0 \,.$$

Using the divergence theorem, the divergence term transforms into the surface integral:

$$\frac{1}{4\pi\varepsilon_0} \int \nabla_{r_0} \cdot \left(p(r_0) \frac{1}{|r - r_0|} \right) d^3 r_0$$

$$= \frac{1}{4\pi\varepsilon_0} \int \frac{p(r_0) \cdot dA_0}{|r - r_0|} \,,$$

with dA_0 an element of surface area of the volume. In the event that $p(r)$ is a constant, only the surface term survives:

$$\phi(r) = \frac{1}{4\pi\varepsilon_0} \int \frac{1}{|r - r_0|} \, p \cdot dA_0 \,,$$

with dA_0 an elementary area of the surface bounding the charges. In words, the potential due to a constant p inside the surface is equivalent to that of a *surface charge:*

$$\sigma = p \cdot dA$$

which is positive for surface elements with a component in the direction of p and negative for surface elements pointed oppositely. (Usually the direction of a surface element is taken to be that of the outward normal to the surface at the location of the element.)

If the bounding surface is a sphere, and the point of observation is at the center of this sphere, the integration over the surface of the sphere is zero: the positive and negative surface charge contributions to the potential cancel. If the point of observation is off-center, however, a net potential can result (depending upon the situation) because the positive and negative charges are at different distances from the point of observation. The field due to the surface charge is:

$$E(r) = -\frac{1}{4\pi\varepsilon_0} \nabla_r \int \frac{1}{|r - r_0|} \, p \cdot dA_0 \,,$$

which, at the center of a spherical bounding surface is not zero (the fields of negative and positive charges on opposite sides of the center add because both fields point the same way) but is instead:

$$E = -\frac{p}{3\varepsilon_0} \,.$$

If we suppose the polarization of the dipoles was induced by an external field, the polarization field opposes the applied field and sometimes is called a depolarization field. In the case when the polarization is outside a spherical cavity, the field in the cavity due to the surrounding dipoles is in the same direction as the polarization.

In particular, if the electric susceptibility is introduced through the approximation:

$$p(r) = \varepsilon_0 \chi(r) E(r) \,,$$

where E, in this case and in the following, represent the external field which induces the polarization. Then:

$$\nabla \cdot p(r) = \nabla \cdot (\chi(r)\varepsilon_0 \, E(r)) = -\rho_b \,.$$

Whenever χ(r) is used to model a step discontinuity at the boundary between two regions, the step produces a surface charge layer. For example, integrating along a normal to the bounding surface from a point just interior to one surface to another point just exterior:

$$\varepsilon_0 \hat{n} \cdot \left[\chi(r_+) E(r_+) - \chi(r_-) E(r_-) \right] = \frac{1}{A_n} \int d\Omega_n \; \rho_b = 0 \;,$$

where A_n, Ω_n indicate the area and volume of an elementary region straddling the boundary between the regions, and $\hat{n}$ a unit normal to the surface. The right side vanishes as the volume shrinks, inasmuch as ρ_b is finite, indicating a discontinuity in E, and therefore a surface charge. That is, where the modeled medium includes a step in permittivity, the polarization density corresponding to the dipole moment density:

$$p(r) = \chi(r) E(r)$$

necessarily includes the contribution of a surface charge.

A physically more realistic modeling of p(r) would have the dipole moment density drop off rapidly, but smoothly to zero at the boundary of the confining region, rather than making a sudden step to zero density. Then the surface charge will not concentrate in an infinitely thin surface, but instead, being the divergence of a smoothly varying dipole moment density, will distribute itself throughout a thin, but finite transition layer.

Dielectric Sphere in Uniform External Electric Field

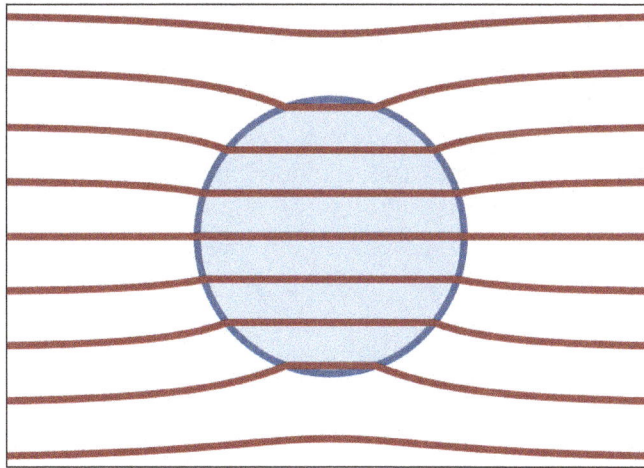

Field lines of the D-field in a dielectric sphere with greater susceptibility than its surroundings, placed in a previously-uniform field. The field lines of the E-field (not shown) coincide everywhere with those of the D-field, but inside the sphere, their density is lower, corresponding to the fact that the E-field is weaker inside the sphere than outside. Many of the external E-field lines terminate on the surface of the sphere, where there is a bound charge.

The above general remarks about surface charge are made more concrete by considering the example of a dielectric sphere in a uniform electric field. The sphere is found to adopt a surface charge related to the dipole moment of its interior.

A uniform external electric field is supposed to point in the z-direction, and spherical-polar coordinates are introduced so the potential created by this field is:

$$\phi_\infty = -E_\infty z = -E_\infty r \cos\theta .$$

The sphere is assumed to be described by a dielectric constant κ, that is,

$$D = \kappa\epsilon_0 E ,$$

and inside the sphere the potential satisfies Laplace's equation. the solution inside the sphere is:

$$\phi_< = Ar\cos\theta ,$$

while outside the sphere:

$$\phi_> = \left(Br + \frac{C}{r^2} \right)\cos\theta .$$

At large distances, $\varphi_> \to \varphi_\infty$ so $B = -E_\infty$. Continuity of potential and of the radial component of displacement $D = \kappa\varepsilon_0 E$ determine the other two constants. Supposing the radius of the sphere is R,

$$A = -\frac{3}{\kappa+2}E_\infty \; ; C = \frac{\kappa-1}{\kappa+2}E_\infty R^3 ,$$

As a consequence, the potential is:

$$\phi_> = \left(-r + \frac{\kappa-1}{\kappa+2}\frac{R^3}{r^2} \right)E_\infty \cos\theta ,$$

which is the potential due to applied field and, in addition, a dipole in the direction of the applied field (the z-direction) of dipole moment:

$$p = 4\pi\varepsilon_0 \left(\frac{\kappa-1}{\kappa+2}R^3 \right)E_\infty ,$$

or, per unit volume:

$$\frac{p}{V} = 3\varepsilon_0 \left(\frac{\kappa-1}{\kappa+2} \right)E_\infty .$$

The factor $(\kappa - 1)/(\kappa + 2)$ is called the Clausius–Mossotti factor and shows that the induced polarization flips sign if $\kappa < 1$. Of course, this cannot happen in this example, but in an example with two different dielectrics κ is replaced by the ratio of the inner to outer region dielectric constants, which can be greater or smaller than one. The potential inside the sphere is:

$$\phi_< = -\frac{3}{\kappa+2}E_\infty r\cos\theta ,$$

leading to the field inside the sphere:

$$-\nabla\phi_< = \frac{3}{\kappa+2}E_\infty = \left(1 - \frac{\kappa-1}{\kappa+2}\right)E_\infty,$$

showing the depolarizing effect of the dipole. Notice that the field inside the sphere is uniform and parallel to the applied field. The dipole moment is uniform throughout the interior of the sphere. The surface charge density on the sphere is the difference between the radial field components:

$$\sigma = 3\varepsilon_0 \frac{\kappa-1}{\kappa+2}E_\infty \cos\theta = \frac{1}{V}p \cdot \hat{R}.$$

This linear dielectric example shows that the dielectric constant treatment is equivalent to the uniform dipole moment model and leads to zero charge everywhere except for the surface charge at the boundary of the sphere.

General Media

If observation is confined to regions sufficiently remote from a system of charges, a multipole expansion of the exact polarization density can be made. By truncating this expansion (for example, retaining only the dipole terms, or only the dipole and quadrupole terms, or *etc.*), In particular, truncating the expansion at the dipole term, the result is indistinguishable from the polarization density generated by a uniform dipole moment confined to the charge region. To the accuracy of this dipole approximation, the dipole moment *density* p(r) (which includes not only p but the location of p) serves as P(r).

At locations *inside* the charge array, to connect an array of paired charges to an approximation involving only a dipole moment density p(r) requires additional considerations. The simplest approximation is to replace the charge array with a model of ideal (infinitesimally spaced) dipoles. In particular, as in the example above that uses a constant dipole moment density confined to a finite region, a surface charge and depolarization field results. A more general version of this model (which allows the polarization to vary with position) is the customary approach using electric susceptibility or electrical permittivity.

A more complex model of the point charge array introduces an effective medium by averaging the microscopic charges; for example, the averaging can arrange that only dipole fields play a role. A related approach is to divide the charges into those nearby the point of observation, and those far enough away to allow a multipole expansion. The nearby charges then give rise to *local field effects*. In a common model of this type, the distant charges are treated as a homogeneous medium using a dielectric constant, and the nearby charges are treated only in a dipole approximation. The approximation of a medium or an array of charges by only dipoles and their associated dipole moment density is sometimes called the *point dipole* approximation, the *discrete dipole approximation*, or simply the *dipole approximation*.

Electric Dipole Moments of Fundamental Particles

Not to be confused with spin which refers to the magnetic dipole moments of particles, much experimental work is continuing on measuring the electric dipole moments (EDM) of fundamental and composite particles, namely those of the electron and neutron, respectively. As EDMs violate both the parity (P) and time-reversal (T) symmetries, their values yield a mostly model-independent measure of CP-violation in nature (assuming CPT symmetry is valid). Therefore, values for these EDMs place strong constraints upon the scale of CP-violation that extensions to the standard model of particle physics may allow. Current generations of experiments are designed to be sensitive to the supersymmetry range of EDMs, providing complementary experiments to those done at the LHC.

Indeed, many theories are inconsistent with the current limits and have effectively been ruled out, and established theory permits a much larger value than these limits, leading to the strong CP problem and prompting searches for new particles such as the axion.

Dipole Moments of Molecules

Dipole moments in molecules are responsible for the behavior of a substance in the presence of external electric fields. The dipoles tend to be aligned to the external field which can be constant or time-dependent. This effect forms the basis of a modern experimental technique called dielectric spectroscopy.

Dipole moments can be found in common molecules such as water and also in biomolecules such as proteins.

By means of the total dipole moment of some material one can compute the dielectric constant which is related to the more intuitive concept of conductivity. If $\mathcal{M}_{Tot}$ is the total dipole moment of the sample, then the dielectric constant is given by,

$$\epsilon = 1 + k \left\langle \mathcal{M}_{Tot}^2 \right\rangle$$

where k is a constant and $\left\langle \mathcal{M}_{Tot}^2 \right\rangle = \left\langle \mathcal{M}_{Tot}(t=0)\mathcal{M}_{Tot}(t=0) \right\rangle$ is the time correlation function of the total dipole moment. In general the total dipole moment have contributions coming from translations and rotations of the molecules in the sample,

$$\mathcal{M}_{Tot} = \mathcal{M}_{Trans} + \mathcal{M}_{Rot}.$$

Therefore, the dielectric constant (and the conductivity) has contributions from both terms. This approach can be generalized to compute the frequency dependent dielectric function.

It is possible to calculate dipole moments from electronic structure theory, either as a response to constant electric fields or from the density matrix. Such values however are not directly comparable to experiment due to the potential presence of nuclear quantum effects, which can be substantial for even simple systems like the ammonia molecule. Coupled cluster theory (especially CCSD(T)) can give very accurate dipole moments, although it is possible to get reasonable estimates (within about 5%) from density functional theory, especially if hybrid or double hybrid functionals are employed. The dipole moment of a molecule can also be calculated based on the molecular structure using the concept of group contribution methods.

Electric Field

An electric field surrounds an electric charge, and exerts force on other charges in the field, attracting or repelling them. Electric field is sometimes abbreviated as E-field. The electric field is defined mathematically as a vector field that associates to each point in space the (electrostatic or Coulomb) force per unit of charge exerted on an infinitesimal positive test charge at rest at that point. The SI unit for electric field strength is volt per meter (V/m). Newtons per coulomb (N/C) is also used as a unit of electric field strength. Electric fields are created by electric charges, or by time-varying magnetic fields. Electric fields are important in many areas of physics, and are exploited practically in electrical technology. On an atomic scale, the electric field is responsible for the attractive force between the atomic nucleus and electrons that holds atoms together, and the forces between atoms that cause chemical bonding. Electric fields and magnetic fields are both manifestations of the electromagnetic force, one of the four fundamental forces (or interactions) of nature.

From Coulomb's law a particle with electric charge q_1 at position x_1 exerts a force on a particle with charge q_0 at position x_0 of

$$F = \frac{1}{4\pi\varepsilon_0} \frac{q_1 q_0}{(x_1 - x_0)^2} \hat{r}_{1,0}$$

where $r_{1,0}$ is the unit vector in the direction from point x_1 to point x_0, and ε_0 is the electric constant (also known as "the absolute permittivity of free space") in $C^2 \, m^{-2} \, N^{-1}$

When the charges q_0 and q_1 have the same sign this force is positive, directed away from the other charge, indicating the particles repel each other. When the charges have unlike signs the force is negative, indicating the particles attract. To make it easy to calculate the Coulomb force on any charge at position x_0 this expression can be divided by q_0, leaving an expression that only depends on the other charge (the *source* charge)

$$E(x_0) = \frac{F}{q_0} = \frac{1}{4\pi\varepsilon_0} \frac{q_1}{(x_1 - x_0)^2} \hat{r}_{1,0}$$

This is the *electric field* at point x_0 due to the point charge q_1; it is a vector equal to the Coulomb force per unit charge that a positive point charge would experience at the position x_0. Since this formula gives the electric field magnitude and direction at any point x_0 in space (except at the location of the charge itself, x_1, where it becomes infinite) it defines a vector field. From the above formula it can be seen that the electric field due to a point charge is everywhere directed away from the charge if it is positive, and toward the charge if it is negative, and its magnitude decreases with the inverse square of the distance from the charge.

If there are multiple charges, the resultant Coulomb force on a charge can be found by summing the vectors of the forces due to each charge. This shows the electric field obeys the *superposition principle*: the total electric field at a point due to a collection of charges is just equal to the vector sum of the electric fields at that point due to the individual charges.

$$E(x) = E_1(x) + E_2(x) + E_3(x) + \cdots = \frac{1}{4\pi\varepsilon_0}\frac{q_1}{(x_1 - x)^2}\hat{r}_1 + \frac{1}{4\pi\varepsilon_0}\frac{q_2}{(x_2 - x)^2}\hat{r}_2 + \frac{1}{4\pi\varepsilon_0}\frac{q_3}{(x_3 - x)^2}\hat{r}_3 + \cdots$$

$$E(x) = \frac{1}{4\pi\varepsilon_0}\sum_{k=1}^{N}\frac{q_k}{(x_k - x)^2}\hat{r}_k$$

where $\hat{r}_k$ is the unit vector in the direction from point x_k to point x.

This is the definition of the electric field due to the point *source charges* $q_1, \ldots, q_N$. It diverges and becomes infinite at the locations of the charges themselves, and so is not defined there.

Evidence of an electric field: styrofoam peanuts clinging to a cat's fur due to static electricity. The triboelectric effect causes an electrostatic charge to build up on the fur due to the cat's motions. The electric field of the charge causes polarization of the molecules of the styrofoam due to electrostatic induction, resulting in a slight attraction of the light plastic pieces to the charged fur. This effect is also the cause of static cling in clothes.

The Coulomb force on a charge of magnitude q at any point in space is equal to the product of the charge and the electric field at that point.

$$F = qE$$

The units of the electric field in the SI system are newtons per coulomb (N/C), or volts per meter (V/m); in terms of the SI base units they are $kg \cdot m \cdot s^{-3} \cdot A^{-1}$.

The electric field due to a continuous distribution of charge $\rho(x)$ in space (where ρ is the charge density in coulombs per cubic meter) can be calculated by considering the charge $\rho(x')\,dV$ in each small volume of space dV at point x' as a point charge, and calculating its electric field $dE(x)$ at point x.

$$dE(x) = \frac{1}{4\pi\varepsilon_0}\frac{\rho(x')dV}{(x' - x)^2}\hat{r}'$$

where r' is the unit vector pointing from x' to x, then adding up the contributions from all the increments of volume by integrating over the volume of the charge distribution V.

$$E(x) = \frac{1}{4\pi\varepsilon_0}\iiint_V \frac{\rho(x')dV}{(x' - x)^2}\hat{r}'$$

Causes and Description

Electric fields are caused by electric charges, described by Gauss's law, or varying magnetic fields, described by Faraday's law of induction. Together, these laws are enough to define the behavior of the electric field as a function of charge repartition and magnetic field. However, since the magnetic field is described as a function of electric field, the equations of both fields are coupled and together form Maxwell's equations that describe both fields as a function of charges and currents.

In the special case of a steady state (stationary charges and currents), the Maxwell-Faraday inductive effect disappears. The resulting two equations (Gauss's law $\nabla \cdot E = \dfrac{\rho}{\varepsilon_0}$ and Faraday's law with no induction term $\nabla \times E = 0$, taken together, are equivalent to Coulomb's law, written as $E(r) = \dfrac{1}{4\pi\varepsilon_0} \int \rho(r') \dfrac{r-r'}{|r-r'|^3} d^3 r'$ for a charge density $\rho(r)$ (r is position in space). Notice that ε_0, the vacuum electric permittivity, must be substituted with ε, permittivity, when charges are in non-empty media.

Continuous vs. Discrete Charge Representation

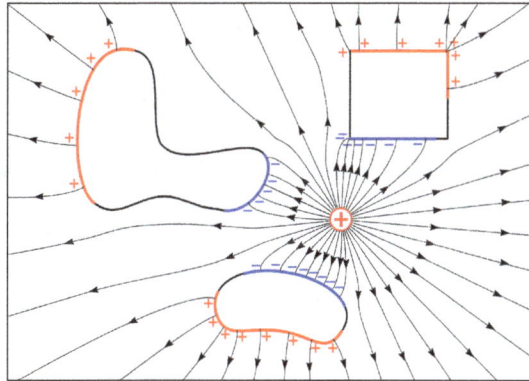

The electric field *(lines with arrows)* of a charge (+) induces surface charges *(red and blue areas)* on metal objects due to electrostatic induction.

The equations of electromagnetism are best described in a continuous description. However, charges are sometimes best described as discrete points; for example, some models may describe electrons as point sources where charge density is infinite on an infinitesimal section of space.

A charge q located at r_0 can be described mathematically as a charge density $\rho(r) = q\delta(r - r_0)$, where the Dirac delta function (in three dimensions) is used. Conversely, a charge distribution can be approximated by many small point charges.

Superposition Principle

Electric fields satisfy the superposition principle, because Maxwell's equations are linear. As a result, if E_1 and E_2 are the electric fields resulting from distribution of charges ρ_1 and ρ_2, a distribution of charges $\rho_1 + \rho_2$ will create an electric field $E_1 + E_2$; for instance, Coulomb's law is linear in charge density as well.

This principle is useful to calculate the field created by multiple point charges. If charges $q_1, q_2, .., q_n$ are stationary in space at $r_1, r_2, . r_n$, in the absence of currents, the superposition principle proves that the resulting field is the sum of fields generated by each particle as described by Coulomb's law:

$$E(r) = \sum_{i=1}^{N} E_i(r) = \frac{1}{4\pi\varepsilon_0} \sum_{i=1}^{N} q_i \frac{r - r_i}{|r - r_i|^3}$$

Electrostatic Fields

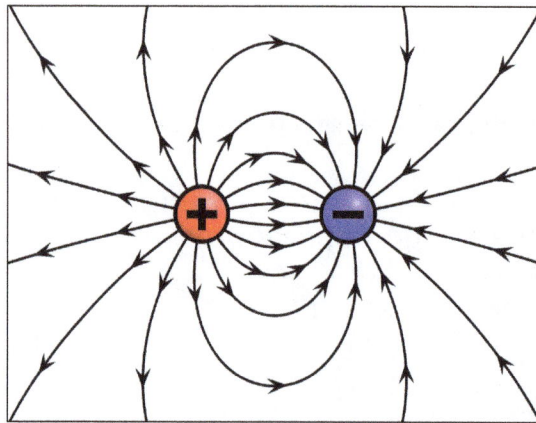

Illustration of the electric field surrounding a positive (red) and a negative (blue) charge.

Experiment illustrating electric field lines. An electrode connected to an electrostatic induction machine is placed in an oil-filled container. Considering that oil is a dielectric medium, when there is current through the electrode, the particles arrange themselves so as to show the force lines of the electric field.

Electrostatic fields are electric fields which do not change with time, which happens when charges and currents are stationary. In that case, Coulomb's law fully describes the field.

Electric Potential

If a system is static, such that magnetic fields are not time-varying, then by Faraday's law, the electric field is curl-free. In this case, one can define an electric potential, that is, a function Φ such that $E = -\nabla\Phi$. This is analogous to the gravitational potential.

Parallels between Electrostatic and Gravitational Fields

Coulomb's law, which describes the interaction of electric charges:

$$F = q \left(\frac{Q}{4\pi\varepsilon_0} \frac{\hat{r}}{|r|^2} \right) = qE$$

is similar to Newton's law of universal gravitation:

$$F = m\left(-GM\frac{\hat{r}}{|r|^2}\right) = mg$$

(where $\hat{r} = \dfrac{r}{|r|}$)

This suggests similarities between the electric field E and the gravitational field g, or their associated potentials. Mass is sometimes called "gravitational charge".

Electrostatic and gravitational forces both are central, conservative and obey an inverse-square law.

Uniform Fields

A uniform field is one in which the electric field is constant at every point. It can be approximated by placing two conducting plates parallel to each other and maintaining a voltage (potential difference) between them; it is only an approximation because of boundary effects (near the edge of the planes, electric field is distorted because the plane does not continue). Assuming infinite planes, the magnitude of the electric field E is:

$$E = -\frac{V}{d}$$

where ΔV is the potential difference between the plates and d is the distance separating the plates. The negative sign arises as positive charges repel, so a positive charge will experience a force away from the positively charged plate, in the opposite direction to that in which the voltage increases. In micro- and nano-applications, for instance in relation to semiconductors, a typical magnitude of an electric field is in the order of 10^6 V·m^{-1}, achieved by applying a voltage of the order of 1 volt between conductors spaced 1 μm apart.

Electrodynamic Fields

Electrodynamic fields are electric fields which do change with time, for instance when charges are in motion.

The electric field cannot be described independently of the magnetic field in that case. If A is the magnetic vector potential, defined so that $B = \nabla \times A$, one can still define an electric potential Φ such that:

$$E = -\nabla\Phi - \frac{\partial A}{\partial t}$$

One can recover Faraday's law of induction by taking the curl of that equation,

$$\nabla \times E = -\frac{\partial(\nabla \times A)}{\partial t} = -\frac{\partial B}{\partial t}$$

which justifies, a posteriori, the previous form for E.

Energy in the Electric Field

The total energy per unit volume stored by the electromagnetic field is:

$$u_{EM} = \frac{\varepsilon}{2} |E|^2 + \frac{1}{2\mu} |B|^2$$

where ε is the permittivity of the medium in which the field exists, μ its magnetic permeability, and E and B are the electric and magnetic field vectors.

As E and B fields are coupled, it would be misleading to split this expression into "electric" and "magnetic" contributions. However, in the steady-state case, the fields are no longer coupled. It makes sense in that case to compute the electrostatic energy per unit volume:

$$u_{ES} = \frac{1}{2} \varepsilon |E|^2,$$

The total energy U stored in the electric field in a given volume V is therefore,

$$U_{ES} = \frac{1}{2} \varepsilon \int_V |E|^2 \, dV,$$

Further Extensions

Definitive Equation of Vector Fields

In the presence of matter, it is helpful to extend the notion of the electric field into three vector fields:

$$D = \varepsilon_0 E + P$$

where P is the electric polarization – the volume density of electric dipole moments, and D is the electric displacement field. Since E and P are defined separately, this equation can be used to define D. The physical interpretation of D is not as clear as E (effectively the field applied to the material) or P (induced field due to the dipoles in the material), but still serves as a convenient mathematical simplification, since Maxwell's equations can be simplified in terms of free charges and currents.

Constitutive Relation

The E and D fields are related by the permittivity of the material, ε.

For linear, homogeneous, isotropic materials E and D are proportional and constant throughout the region, there is no position dependence: For inhomogeneous materials, there is a position dependence throughout the material:

$$D(r) = \varepsilon E(r)$$

For anisotropic materials the E and D fields are not parallel, and so E and D are related by the permittivity tensor (a 2nd order tensor field), in component form:

$$D_i = \varepsilon_{ij} E_j$$

For non-linear media, E and D are not proportional. Materials can have varying extents of linearity, homogeneity and isotropy.

Electric Flux

Electric flux is a property of an electric field that may be thought of as the number of electric lines of force (or electric field lines) that intersect a given area. Electric field lines are considered to originate on positive electric charges and to terminate on negative charges. Field lines directed into a closed surface are considered negative; those directed out of a closed surface are positive. If there is no net charge within a closed surface, every field line directed into the surface continues through the interior and is directed outward elsewhere on the surface. The negative flux just equals in magnitude the positive flux, so that the net, or total, electric flux is zero. If a net charge is contained inside a closed surface, the total flux through the surface is proportional to the enclosed charge, positive if it is positive, negative if it is negative.

The mathematical relation between electric flux and enclosed charge is known as Gauss's law for the electric field, one of the fundamental laws of electromagnetism. In the metre-kilogram-second system and the International System of Units (SI) the net flux of an electric field through any closed surface is equal to the enclosed charge, in units of coulombs, divided by a constant, called the permittivity of free space; in the centimetre-gram-second system the net flux of an electric field through any closed surface is equal to the constant 4π times the enclosed charge, in electrostatic units (esu).

Gauss's Law

In physics, Gauss's law, also known as Gauss's flux theorem, is a law relating the distribution of electric charge to the resulting electric field. The surface under consideration may be a closed one enclosing a volume such as a spherical surface.

The law was first formulated by Joseph-Louis Lagrange in 1773, followed by Carl Friedrich Gauss in 1813, both in the context of the attraction of ellipsoids. It is one of Maxwell's four equations, which form the basis of classical electrodynamics. Gauss's law can be used to derive Coulomb's law, and vice versa.

Qualitative Description

In words, Gauss's law states that:

The net electric flux through any hypothetical closed surface is equal to $\frac{1}{\varepsilon_0}$ times the net electric charge within that closed surface.

Gauss's law has a close mathematical similarity with a number of laws in other areas of physics, such as Gauss's law for magnetism and Gauss's law for gravity. In fact, any inverse-square law can be formulated in a way similar to Gauss's law: for example, Gauss's law itself is essentially equivalent to the inverse-square Coulomb's law, and Gauss's law for gravity is essentially equivalent to the inverse-square Newton's law of gravity.

The law can be expressed mathematically using vector calculus in integral form and differential form; both are equivalent since they are related by the divergence theorem, also called Gauss's theorem. Each of these forms in turn can also be expressed two ways: In terms of a relation between the electric field E and the total electric charge, or in terms of the electric displacement field D and the free electric charge.

Equation Involving the E Field

Gauss's law can be stated using either the electric field E or the electric displacement field D.

Integral Form

Gauss's law may be expressed as:

$$\Phi_E = \frac{Q}{\varepsilon_0}$$

where Φ_E is the electric flux through a closed surface S enclosing any volume V, Q is the total charge enclosed within V, and ε_0 is the electric constant. The electric flux Φ_E is defined as a surface integral of the electric field:

$$\Phi_E = \oiint_S E \cdot dA$$

where E is the electric field, dA is a vector representing an infinitesimal element of area of the surface, and · represents the dot product of two vectors.

Since the flux is defined as an *integral* of the electric field, this expression of Gauss's law is called the *integral form*.

An important fact about this fundamental equation often doesn't find a mention in expositions that are not absolutely diligent. The above equation may fail to hold true in case the closed surface S contains a singularity of the electric field, which is physicists' term for a point in space where either a point charge exists and the field strength approaches infinity, or the field's magnitude or direction gets altered discontinuously due to the existence of a surface charge. In 2011, a modification of the above equation, called the Generalized Gauss's Theorem by its original creator, was published in the proceedings of the 2011 Annual Meeting of Electrostatics Society of America. The Generalized Gauss's Theorem allows the closed surface S to pass through singularities of the electric field. A corollary of the Generalized Gauss's Theorem, known as the simplest form of the Generalized Gauss's Theorem, holds true if the surface S is smooth. It states that:

$$\Phi_E = \frac{Q}{\varepsilon_0} + \frac{1}{2}\frac{Q'}{\varepsilon_0}$$

where Q is the net charge enclosed within V and Q' is the net charge contained by the closed surface S itself.

Applying the Integral Form

If the electric field is known everywhere, Gauss's law makes it possible to find the distribution of electric charge: The charge in any given region can be deduced by integrating the electric field to find the flux.

The reverse problem (when the electric charge distribution is known and the electric field must be computed) is much more difficult. The total flux through a given surface gives little information about the electric field, and can go in and out of the surface in arbitrarily complicated patterns.

An exception is if there is some symmetry in the problem, which mandates that the electric field passes through the surface in a uniform way. Then, if the total flux is known, the field itself can be deduced at every point. Common examples of symmetries which lend themselves to Gauss's law include: cylindrical symmetry, planar symmetry, and spherical symmetry.

Differential Form

By the divergence theorem, Gauss's law can alternatively be written in the differential form:

$$\nabla \cdot E = \frac{\rho}{\varepsilon_0}$$

where $\nabla \cdot E$ is the divergence of the electric field, ε_0 is the electric constant, and ρ is the total electric charge density (charge per unit volume).

Equivalence of Integral and Differential Forms

The integral and differential forms are mathematically equivalent, by the divergence theorem. Here is the argument more specifically.

Proof

The integral form of Gauss' law is:

$$\oiint_S E \cdot dA = \frac{Q}{\varepsilon_0}$$

for any closed surface S containing charge Q. By the divergence theorem, this equation is equivalent to:

$$\iiint_V \nabla \cdot E \, dV = \frac{Q}{\varepsilon_0}$$

for any volume V containing charge Q. By the relation between charge and charge density, this equation is equivalent to:

$$\iiint_V \nabla \cdot E \, dV = \iiint_V \frac{\rho}{\varepsilon_0} \, dV$$

for any volume V. In order for this equation to be simultaneously true for every possible volume V, it is necessary (and sufficient) for the integrands to be equal everywhere. Therefore, this equation is equivalent to:

$$\nabla \cdot E = \frac{\rho}{\varepsilon_0}.$$

Thus the integral and differential forms are equivalent.

Equation Involving the D Field

Free, Bound and Total Charge

The electric charge that arises in the simplest textbook situations would be classified as "free charge"— for example, the charge which is transferred in static electricity, or the charge on a capacitor plate. In contrast, "bound charge" arises only in the context of dielectric (polarizable) materials. (All materials are polarizable to some extent.) When such materials are placed in an external electric field, the electrons remain bound to their respective atoms, but shift a microscopic distance in response to the field, so that they're more on one side of the atom than the other. All these microscopic displacements add up to give a macroscopic net charge distribution, and this constitutes the "bound charge".

Although microscopically all charge is fundamentally the same, there are often practical reasons for wanting to treat bound charge differently from free charge. The result is that the more fundamental Gauss's law, in terms of E (above), is sometimes put into the equivalent form below, which is in terms of D and the free charge only.

Integral Form

This formulation of Gauss's law states the total charge form:

$$\Phi_D = Q_{free}$$

where Φ_D is the D-field flux through a surface S which encloses a volume V, and Q_{free} is the free charge contained in V. The flux Φ_D is defined analogously to the flux Φ_E of the electric field E through S:

$$\Phi_D = \oiint_S D \cdot dA$$

Differential Form

The differential form of Gauss's law, involving free charge only, states:

$$\nabla \cdot D = \rho_{free}$$

where $\nabla \cdot D$ is the divergence of the electric displacement field, and ρ_{free} is the free electric charge density.

Equivalence of Total and Free Charge Statements

Proof that the formulations of Gauss's law in terms of free charge are equivalent to the formulations involving total charge.

In this proof, we will show that the equation,

$$\nabla \cdot E = \frac{\rho}{\varepsilon_0}$$

is equivalent to the equation,

$$\nabla \cdot D = \rho_{\text{free}}$$

Note that we are only dealing with the differential forms, not the integral forms, but that is sufficient since the differential and integral forms are equivalent in each case, by the divergence theorem.

We introduce the polarization density P, which has the following relation to E and D:

$$D = \varepsilon_0 E + P$$

and the following relation to the bound charge:

$$\rho_{bound} = -\nabla \cdot P$$

Now, consider the three equations:

$$\rho_{bound} = \nabla \cdot (-P)$$
$$\rho_{free} = \nabla \cdot D$$
$$\rho = \nabla \cdot (\varepsilon_0 E)$$

The key insight is that the sum of the first two equations is the third equation. This completes the proof: The first equation is true by definition, and therefore the second equation is true if and only if the third equation is true. So the second and third equations are equivalent, which is what we wanted to prove.

Equation for Linear Materials

In homogeneous, isotropic, nondispersive, linear materials, there is a simple relationship between E and D:

$$D = \varepsilon E$$

where ε is the permittivity of the material. For the case of vacuum (aka free space), $\varepsilon = \varepsilon_0$. Under these circumstances, Gauss's law modifies to,

$$\Phi_E = \frac{Q_{free}}{\varepsilon}$$

for the integral form, and

$$\nabla \cdot E = \frac{\rho_{free}}{\varepsilon}$$

for the differential form.

Interpretations

In Terms of Fields of Force

Gauss's theorem can be interpreted in terms of the lines of force of the field as follows:

The flux through a closed surface is dependent upon both the magnitude and direction of the electric field lines penetrating the surface. In general a positive flux is defined by these lines leaving the surface and negative flux by lines entering this surface. This results in positive charges causing a positive flux and negative charges creating a negative flux. These electric field lines will extend to infinite decreasing in strength by a factor of one over the distance from the source of the charge squared. The larger the number of field lines emanating from a charge the larger the magnitude of the charge is, and the closer together the field lines are the greater the magnitude of the electric field. This has the natural result of the electric field becoming weaker as one moves away from a charged particle, but the surface area also increases so that the net electric field exiting this particle will stay the same. In other words the closed integral of the electric field and the dot product of the derivative of the area will equal the net charge enclosed divided by permittivity of free space.

Relation to Coulomb's Law

Deriving Gauss's Law From Coulomb's Law

Strictly speaking, Gauss's law cannot be derived from Coulomb's law alone, since Coulomb's law gives the electric field due to an individual point charge only. However, Gauss's law *can* be proven from Coulomb's law if it is assumed, in addition, that the electric field obeys the superposition principle. The superposition principle says that the resulting field is the vector sum of fields generated by each particle (or the integral, if the charges are distributed smoothly in space).

Proof: Coulomb's law states that the electric field due to a stationary point charge is:

$$E(r) = \frac{q}{4\pi\varepsilon_0} \frac{e_r}{r^2}$$

where,

      e_r is the radial unit vector,

      r is the radius, $|\mathbf{r}|$,

ε_0 is the electric constant,

q is the charge of the particle, which is assumed to be located at the origin.

Using the expression from Coulomb's law, we get the total field at r by using an integral to sum the field at r due to the infinitesimal charge at each other point s in space, to give:

$$E(r) = \frac{1}{4\pi\varepsilon_0} \int \frac{\rho(s)(r-s)}{|r-s|^3} d^3s$$

where ρ is the charge density. If we take the divergence of both sides of this equation with respect to r, and use the known theorem:

$$\nabla \cdot \left(\frac{r}{|r|^3} \right) = 4\pi\delta(r)$$

where $\delta(r)$ is the Dirac delta function, the result is,

$$\nabla \cdot E(r) = \frac{1}{\varepsilon_0} \int \rho(s)\delta(r-s) d^3s$$

Using the "sifting property" of the Dirac delta function, we arrive at,

$$\nabla \cdot E(r) = \frac{\rho(r)}{\varepsilon_0},$$

which is the differential form of Gauss' law, as desired.

Note that since Coulomb's law only applies to stationary charges, there is no reason to expect Gauss's law to hold for moving charges based on this derivation alone. In fact, Gauss's law does hold for moving charges, and in this respect Gauss's law is more general than Coulomb's law.

Deriving Coulomb's Law from Gauss's Law

Strictly speaking, Coulomb's law cannot be derived from Gauss's law alone, since Gauss's law does not give any information regarding the curl of E. However, Coulomb's law *can* be proven from Gauss's law if it is assumed, in addition, that the electric field from a point charge is spherically symmetric (this assumption, like Coulomb's law itself, is exactly true if the charge is stationary, and approximately true if the charge is in motion).

Proof: Taking S in the integral form of Gauss' law to be a spherical surface of radius r, centered at the point charge Q, we have,

$$\oint_S E \cdot dA = \frac{Q}{\varepsilon_0}$$

By the assumption of spherical symmetry, the integrand is a constant which can be taken out of the integral. The result is,

$$4\pi r^2 \hat{r} \cdot E(r) = \frac{Q}{\varepsilon_0}$$

where $\hat{r}$ is a unit vector pointing radially away from the charge. Again by spherical symmetry, E points in the radial direction, and so we get:

$$E(r) = \frac{Q}{4\pi\varepsilon_0}\frac{\hat{r}}{r^2}$$

which is essentially equivalent to Coulomb's law. Thus the inverse-square law dependence of the electric field in Coulomb's law follows from Gauss' law.

Gaussian Surface

A Gaussian surface (sometimes abbreviated as G.S.) is a closed surface in three-dimensional space through which the flux of a vector field is calculated; usually the gravitational field, the electric field, or magnetic field. It is an arbitrary closed surface $S = \partial V$ (the boundary of a 3-dimensional region V) used in conjunction with Gauss's law for the corresponding field (Gauss's law, Gauss's law for magnetism, or Gauss's law for gravity) by performing a surface integral, in order to calculate the total amount of the source quantity enclosed; e.g., amount of gravitational mass as the source of the gravitational field or amount of electric charge as the source of the electrostatic field, or vice versa: calculate the fields for the source distribution.

For concreteness, the electric field is considered in this article, as this is the most frequent type of field the surface concept is used for.

Gaussian surfaces are usually carefully chosen to exploit symmetries of a situation to simplify the calculation of the surface integral. If the Gaussian surface is chosen such that for every point on the surface the component of the electric field along the normal vector is constant, then the calculation will not require difficult integration as the constants which arise can be taken out of the integral. It is defined as the closed surface in three dimensional space by which the flux of vector field be calculated.

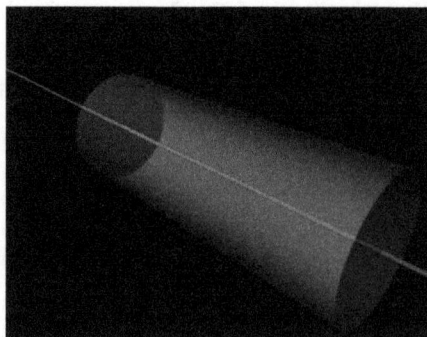

A cylindrical Gaussian surface is commonly used to calculate the electric charge of an infinitely long, straight, 'ideal' wire.

Common Gaussian Surfaces

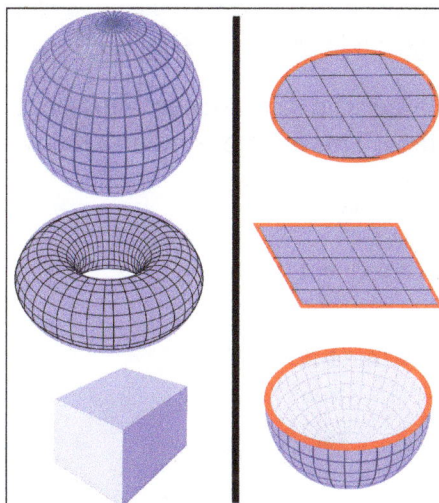

Examples of valid (left) and invalid (right) Gaussian surfaces. Left: Some valid Gaussian surfaces include the surface of a sphere, surface of a torus, and surface of a cube. They are closed surfaces that fully enclose a 3D volume. Right: Some surfaces that CANNOT be used as Gaussian surfaces, such as the disk surface, square surface, or hemisphere surface. They do not fully enclose a 3D volume, and have boundaries (red). Note that infinite planes can approximate Gaussian surfaces.

Most calculations using Gaussian surfaces begin by implementing Gauss's law (for electricity):

$$\Phi_E = \oiint_S E \cdot dA = \frac{Q_{enc}}{\varepsilon_0}$$

Thereby Q_{enc} is the electrical charge enclosed by the Gaussian surface.

This is Gauss's law, combining both the divergence theorem and Coulomb's law.

Spherical Surface

A spherical Gaussian surface is used when finding the electric field or the flux produced by any of the following:

- A point charge.

- A uniformly distributed spherical shell of charge.

- Any other charge distribution with spherical symmetry.

The spherical Gaussian surface is chosen so that it is concentric with the charge distribution.

As an example, consider a charged spherical shell S of negligible thickness, with a uniformly distributed charge Q and radius R. We can use Gauss's law to find the magnitude of the resultant electric field E at a distance r from the center of the charged shell. It is immediately apparent that for a spherical Gaussian surface of radius $r < R$ the enclosed charge is zero: hence the net flux is zero and the magnitude of the electric field on the Gaussian surface is also o (by letting $Q_A = $ o in Gauss's law, where Q_A is the charge enclosed by the Gaussian surface).

With the same example, using a larger Gaussian surface outside the shell where $r > R$, Gauss's law will produce a non-zero electric field. This is determined as follows.

The flux out of the spherical surface S is:

$$\Phi_E = \oiint_{\partial S} E \cdot dA = \int\int_c EdA \cos 0° = E \int\int_S dA$$

The surface area of the sphere of radius r is,

$$\int\int_S dA = 4\pi r^2$$

which implies,

$$\Phi_E = E4\pi r^2$$

By Gauss's law the flux is also,

$$\Phi_E = \frac{Q_A}{\varepsilon_0}$$

finally equating the expression for Φ_E gives the magnitude of the E-field at position r:

$$E4\pi r^2 = \frac{Q_A}{\varepsilon_0} \quad \Rightarrow \quad E = \frac{Q_A}{4\pi\varepsilon_0 r^2}.$$

This non-trivial result shows that any spherical distribution of charge *acts as a point charge* when observed from the outside of the charge distribution; this is in fact a verification of Coulomb's law. And, any exterior charges do not count.

Cylindrical Surface

A cylindrical Gaussian surface is used when finding the electric field or the flux produced by any of the following:

- An infinitely long line of uniform charge.
- An infinite plane of uniform charge.
- An infinitely long cylinder of uniform charge.

As example "field near infinite line charge" is given below:

Consider a point P at a distance r from an infinite line charge having charge density (charge per unit length) λ. Imagine a closed surface in the form of cylinder whose axis of rotation is the line charge. If h is the length of the cylinder, then the charge enclosed in the cylinder is,

$$q = \lambda h,$$

where q is the charge enclosed in the Gaussian surface. There are three surfaces a, b and c as shown in the figure. The differential vector area is dA, on each surface a, b and c.

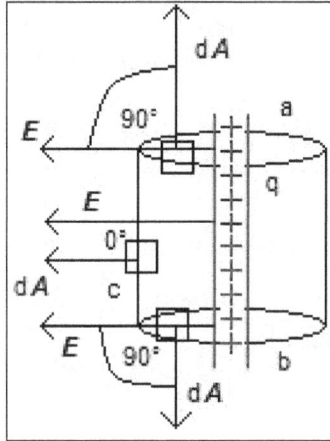

Closed surface in the form of a cylinder having line charge in the center and showing differential areas dA of all three surfaces.

The flux passing consists of the three contributions:

$$\Phi_E = \oiint_A E \cdot dA = \int\int_a E \cdot dA + \int\int_b E \cdot dA + \int\int_c E \cdot dA$$

For surfaces a and b, E and dA will be perpendicular. For surface c, E and dA will be parallel, as shown in the figure.

$$\Phi_E = \int\int_a EdA\cos 90^\circ + \int\int_b EdA\cos 90^\circ + \int\int_c EdA\cos 0^\circ$$
$$= E\int\int_c dA$$

The surface area of the cylinder is,

$$\int\int_c dA = 2\pi rh$$

which implies,

$$\Phi_E = E2\pi rh$$

By Gauss's law,

$$\Phi_E = \frac{q}{\varepsilon_0}$$

equating for Φ_E yields,

$$E2\pi rh = \frac{\lambda h}{\varepsilon_0} \quad \Rightarrow \quad E = \frac{\lambda}{2\pi\varepsilon_0 r}$$

Gaussian Pillbox

This surface is most often used to determine the electric field due to an infinite sheet of charge with uniform charge density, or a slab of charge with some finite thickness. The pillbox has a cylindrical shape, and can be thought of as consisting of three components: the disk at one end of the cylinder with area πR^2, the disk at the other end with equal area, and the side of the cylinder. The sum of the electric flux through each component of the surface is proportional to the enclosed charge of the pillbox, as dictated by Gauss's Law. Because the field close to the sheet can be approximated as constant, the pillbox is oriented in a way so that the field lines penetrate the disks at the ends of the field at a perpendicular angle and the side of the cylinder are parallel to the field lines.

Electric Potential

Electric potential is the amount of work needed to move a unit charge from a reference point to a specific point against an electric field. Typically, the reference point is Earth, although any point beyond the influence of the electric field charge can be used.

The diagram shows the forces acting on a positive charge q located between two plates, A and B, of an electric field E. The electric force F exerted by the field on the positive charge is F = qE; to move the charge from plate A to plate B, an equal and opposite force (F′ = −qE) must then be applied. The work W done in moving the positive charge through a distance d is W = F′d = −qEd.

The potential energy for a positive charge increases when it moves against an electric field and decreases when it moves with the electric field; the opposite is true for a negative charge. Unless the unit charge crosses a changing magnetic field, its potential at any given point does not depend on the path taken.

Although the concept of electric potential is useful in understanding electrical phenomena, only differences in potential energy are measurable. If an electric field is defined as the force per unit charge, then by analogy an electric potential can be thought of as the potential energy per unit charge. Therefore, the work done in moving a unit charge from one point to another (e.g., within an electric circuit) is equal to the difference in potential energies at each point. In the International System of Units (SI), electric potential is expressed in units of joules per coulomb (i.e., volts), and differences in potential energy are measured with a voltmeter.

The unit of electric potential is volt. To bring a unit charge from one point to another, if one joule work is done, then the potential difference between the points is said to be one volt. So, we can say,

$$volt = \frac{joules}{coulomb}$$

If one point has electric potential 5 volt, then we can say to bring one coulomb charge from infinity to that point, 5 joule work has to be done.

If one point has potential 5 volt and another point has potential 8 volt, then 8 − 5 or 3 joules work to be done to move one coulomb from first point to second.

Potential at a Point Due to Point Charge

Let us take a positive charge + Q in the space. Let us imagine a point at a distance x from the said charge + Q. Now we place a unit positive charge at that point. As per Coulomb's law, the unit positive charge will experience a force,

$$F = \frac{Q}{4\pi\varepsilon_0\varepsilon_r x^2}$$

Now, let us move this unit positive charge, by a small distance dx towards charge Q.

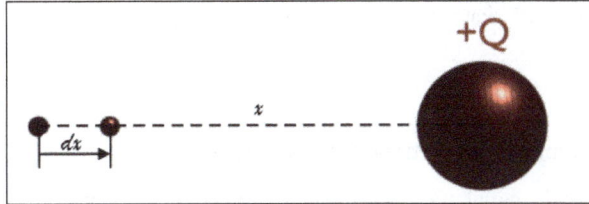

During this movement the work done against the field is,

$$dw = -F \cdot dx = -\frac{Q}{4\pi\varepsilon_0\varepsilon_r x^2}$$

So, total work to be done for bringing the positive unit charge from infinity to distance x, is given by,

$$-\int_\infty^x dw = -\int_0^x \frac{Q}{4\pi\varepsilon_0\varepsilon_r x^2} \cdot dx = \frac{Q}{4\pi\varepsilon_0\varepsilon_r}\left[\frac{1}{x}\right]_\infty^x = \frac{Q}{4\pi\varepsilon_0\varepsilon_r}\left[\frac{1}{x} - \frac{1}{\infty}\right] = \frac{Q}{4\pi\varepsilon_0\varepsilon_r x}$$

As per definition, this is the electric potential of the point due to charge + Q. So, we can write,

$$V = \frac{Q}{4\pi\varepsilon_0\varepsilon_r x}$$

Potential Difference between Two Points

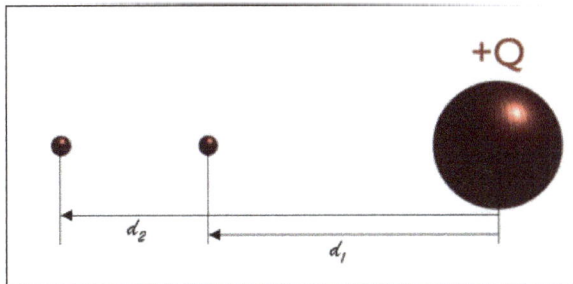

Let us consider two points at distance d_1 meter and d_2 meter from a charge +Q.

We can express the electric potential at the point d_1 meter away from $+Q$, as,

$$Vd_1 = \frac{Q}{4\pi\varepsilon_0\varepsilon_r d_1}$$

We can express the electric potential at the point d_2 meter away from $+Q$, as,

$$Vd_2 = \frac{Q}{4\pi\varepsilon_0\varepsilon_r d_2}$$

Thus, the potential difference between these two points is,

$$Vd_1 - Vd_2 = \frac{Q}{4\pi\varepsilon_0\varepsilon_r d_1} - \frac{Q}{4\pi\varepsilon_0\varepsilon_r d_2} = \frac{Q}{4\pi\varepsilon_0\varepsilon_r}\left[\frac{1}{d_1} - \frac{1}{d_2}\right].$$

References

- Electricity, science: britannica.com, Retrieved 8 January, 2019

- Gambhir, rs; banerjee, d; durgapal, mc (1993). Foundations of physics, vol. 2. New dehli: wiley eastern limited. P. 51. Isbn 9788122405231. Retrieved 10 october 2018

- Electrical-conductor: electrical4u.com, Retrieved 13 May, 2019

- O'grady, patricia f. (2002). Thales of miletus: the beginnings of western science and philosophy. Ashgate. P. 8. Isbn 978-1351895378

- Electrostatics-of-conductors, electrostatic-potential-and-capacitance, physics, guides: toppr.com, Retrieved 25 February, 2019

- 2018 codata value: elementary charge". The nist reference on constants, units, and uncertainty. Nist. 20 may 2019. Retrieved 2019-05-20

- Walker, jearl; halliday, david; resnick, robert (2014). Fundamentals of physics (10th ed.). Hoboken, nj: wiley. Isbn 9781118230732. Oclc 950235056

- Electric-flux, science: britannica.com, Retrieved 16 January, 2019

3

Magnetostatics

The study of magnetic fields in the systems where the currents are steady is known as magnetostatics. Some of the important concepts within this field are Gauss's law for magnetism, Ampère's circuital law and magnetic dipole moment. This chapter closely examines these key concepts of magnetostatics.

Magnetostatics is a branch of electromagnetic studies involving magnetic felds produced by steady non-time varying currents. Evidently currents are produced by moving charges undergoing translational motion. An effective current (called magnetization current) is also produced if magnetic dipoles are nonuniformly distributed. To realize steady currents, a large number of charges must be involved so that collection of charges can be regarded as a continuous fluid. In non-time varying cases ($\partial = \partial t = 0$), the charge conservation principle requires that,

$$\frac{\partial \rho}{\partial t} + \nabla \cdot J = \nabla \cdot J = 0, \text{ in steady state.}$$

This implies that a steady current is transverse. A static magnetic field can also be produced by a nonuniform magnetization (magnetic dipole density in the case of a collection of magnetic dipoles) M (A/m) which produces a magnetization current,

$$J_M = \nabla \times M, \left(A\, m^{-2} \right)$$

The magnetization current is transverse (divergence-free), since $\nabla \cdot J_M = \nabla \cdot \left(\nabla \times M \right) \equiv 0$.

The magnetic field B exerts a force perpendicular to the velocity of charged particles. The force to act on a charge q is.

$$F = q \left(E + v \times B \right)$$

The magnetic field does not do any work on charged particles since the rate of work $v \cdot F$ is independent of B,

(Note that $v \cdot (v \times B) = 0$:) In contrast to the electric field which is a true (or polar) vector, the magnetic field is a pseudo (or axial) vector, or more precisely, constitutes a pseudotensor,

$$B^{ij}\begin{bmatrix} 0 & -B_z & B_y \\ B_z & 0 & -B_x \\ -B_y & B_x & 0 \end{bmatrix}$$

The magnetic field due to a prescribed current distribution can be calculated using the BiotSavartís law. This law is generic because all other laws in magnetostatics, such as the Ampereís law and vanishing divergence of the magnetic field $\nabla \cdot B = 0$; follow from it. Calculation of the magnetic field is facilitated by introducing a magnetic vector potential A; which is related to the magnetic field through $B = \nabla \times A$: This expression is consistent with $\nabla \cdot B = 0$; since $\nabla \cdot (\nabla \times A)$ o identically. It also follows that $\nabla \cdot A = 0$ in static cases.

One important objective of magnetostatics is to derive formulae for the self-inductance and mutual inductance for given current configurations. Use of the vector potential greatly facilitates calculation of the magnetic flux.

$$\Psi = \int B \cdot dS = \int \nabla \times A \cdot dS = \oint A \cdot dl,$$

and inductance,

$$L = \frac{\Psi}{I}.$$

Biot–Savart Law

The Biot Savart Law is an equation describing the magnetic field generated by a constant electric current. It relates the magnetic field to the magnitude, direction, length, and proximity of the electric current. Biot–Savart law is consistent with both Ampere's circuital law and Gauss's theorem. The Biot Savart law is fundamental to magnetostatics, playing a role similar to that of Coulomb's law in electrostatics.

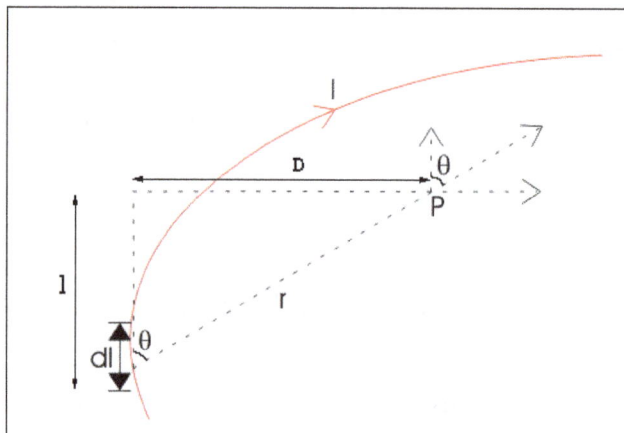

Biot Savart law was created by two French physicists, Jean Baptiste Biot and Felix Savart derived the mathematical expression for magnetic flux density at a point due a nearby current carrying conductor, in 1820. Viewing the deflection of a magantic compass needle, these two scientists concluded that any current element projects a magnetic field into the space around it.

Through observations and calculations they had derived a mathematical expression, which shows, the magnetic flux density of which dB, is directly proportional to the length of the element dl, the current I, the sine of the angle and θ between direction of the current and the vector joining a given point of the field and the current element and is inversely proportional to the square of the distance of the given point from the current element, r.

Biot Savart Law Statement

This is Biot Savart law statement:

$$\text{Hence,} \quad dB \propto \frac{Idl \, \sin\theta}{r^2} \quad \text{or} \quad dB = \frac{I \, dl \sin\theta}{r^2}$$

Where, k is a constant, depending upon the magnetic properties of the medium and system of the units employed. In SI system of unit,

$$k = \frac{\mu_o \mu_r}{4\pi}$$

Therefore, final Biot Savart law derivation is,

$$dB = \frac{\mu_o \mu_r}{4\pi} \times \frac{I \, dl \sin\theta}{r^2}$$

Let us consider a long wire carrying a current I and also consider a point p in the space. The wire is presented in the picture below, by red color. Let us also consider an infinitely small length of the wire dl at a distance r from the point P as shown. Here, r is a distance vector which makes an angle θ with the direction of current in the infinitesimal portion of the wire.

If you try to visualize the condition, you can easily understand the magnetic field density at the point P due to that infinitesimal length dl of the wire is directly proportional to current carried by this portion of the wire.

As the current through that infinitesimal length of wire is same as the current carried by the whole wire itself, we can write,

$$dB \propto I$$

It is also very natural to think that the magnetic field density at that point P due to that infinitesimal length dl of wire is inversely proportional to the square of the straight distance from point P to center of dl. Mathematically we can write this as,

$$dB \propto \frac{1}{r^2}$$

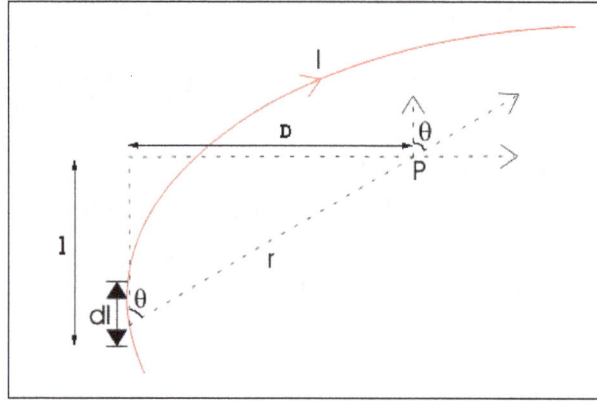

Lastly, magnetic field density at that point P due to that infinitesimal portion of wire is also directly proportional to the actual length of the infinitesimal length dl of wire. As θ be the angle between distance vector r and direction of current through this infinitesimal portion of the wire, the component of dl directly facing perpendicular to the point P is dlsinθ,

Hence, $dB \propto dl \sin \theta$

Now, combining these three statements, we can write,

$$dB \propto \frac{I \cdot dl \sin \theta}{r^2}$$

This is the basic form of Biot Savart's Law.

Now, putting the value of constant k in the above expression, we get,

$$dB \propto \frac{I \cdot dl \sin \theta}{r^2}$$

$$\Rightarrow dB = \frac{\mu_o \mu_r}{4\pi} \times \frac{I \, dl \sin \theta}{r^2}$$

Here, μ_o used in the expression of constant k is absolute permeability of air or vacuum and it's value is $4\pi 10^{-7} W_b$/ A-m in SI system of units. μ_r of the expression of constant k is relative permeability of the medium.

Now, flux density(B) at the point P due to total length of the current carrying conductor or wire can be represented as,

$$B = \int dB \Rightarrow dB = \int \frac{\mu_o \mu_r}{4\pi} \times \frac{I\, dl \sin \theta}{r^2} = \frac{I\mu_o\mu_r}{4\pi} \int \frac{\sin \theta}{r^2} dl$$

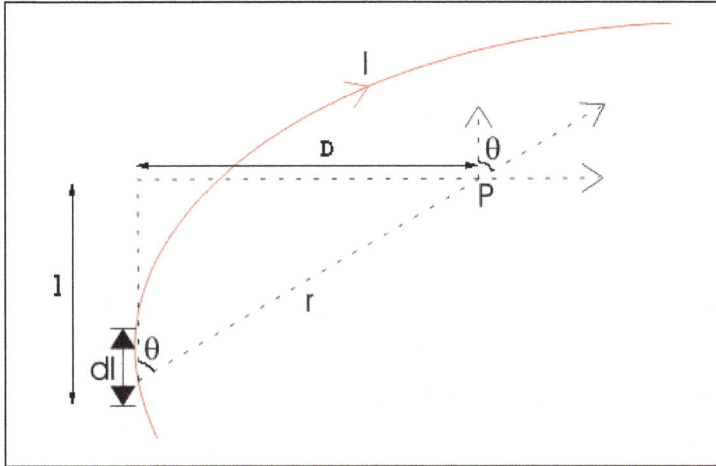

If D is the perpendicular distance of the point P form the wire, then:

$$r \sin \theta = D \ \ or \ r = \frac{D}{\sin \theta}$$

Now, the expression of flux density B at point P can be rewritten as,

$$B = \frac{I \mu_o \mu_r}{4\pi} \int \frac{\sin \theta}{r^2} dl = \frac{I\mu_o\mu_r}{4\pi} \int \frac{\sin^3 \theta}{D^2} dl$$

$$Again, \ \frac{l}{D} = \cot \theta \Rightarrow l = D \cot \theta$$

As per the figure above,

$$Therefore, \ dl = -D \csc^2 \theta\, d\theta$$

Finally the expression of B comes as,

$$B = \frac{I \mu_o \mu_r}{4\pi} \int \frac{\sin^3 \theta}{D^2} \left[-D \csc^2 \theta\, d\theta \right]$$

$$= -\frac{I \mu_o \mu_r}{4\pi} \int \sin^3 \theta \csc^2 \theta\, d\theta$$

$$= -\frac{I \mu_o \mu_r}{4\pi D} \int \sin \theta\, d\theta$$

This angle θ depends upon the length of the wire and the position of the point P. Say for certain limited length of the wire, angle θ as indicated in the figure above varies from θ_1 to θ_2. Hence, magnetic flux density at point P due to total length of the conductor is,

$$B = -\frac{I\,\mu_o\mu_r}{4\pi}\int_{\theta_1}^{\theta_2}\sin\theta\,d\theta$$

$$= -\frac{I\,\mu_o\mu_r}{4\pi D}\left[-\cos\theta\right]_{\theta_1}^{\theta_2}$$

$$= \frac{I\,\mu_o\mu_r}{4\pi D}\left[\cos\theta - \cos\theta_2\right]$$

Let's imagine the wire is infinitely long, then θ will vary from 0 to π that is $\theta_1 = 0$ to $\theta_2 = \pi$. Putting these two values in the above final expression of Biot Savart law, we get,

$$B = \frac{I\,\mu_o\mu_r}{4\pi D}\left[\cos 0 - \cos\pi\right] = B = -\frac{I\,\mu_o\mu_r}{4\pi D}\left[1-(-1)\right] = B = -\frac{\mu_o\mu_r}{2\pi D}I$$

This is nothing but the expression of Ampere's Law.

Magnetic Field

A magnetic field is a vector field that describes the magnetic influence of electric charges in relative motion and magnetized materials. The effects of magnetic fields are commonly seen in permanent magnets, which pull on magnetic materials (such as iron) and attract or repel other magnets. Magnetic fields surround and are created by magnetized material and by moving electric charges (electric currents) such as those used in electromagnets. They exert forces on nearby moving electrical charges and torques on nearby magnets. In addition, a magnetic field that varies with location exerts a force on magnetic materials. Both the strength and direction of a magnetic field vary with location. As such, it is described mathematically as a vector field.

In electromagnetics, the term "magnetic field" is used for two distinct but closely related fields denoted by the symbols B and H. In the International System of Units, H, magnetic field strength, is measured in the SI base units of ampere per meter. B, magnetic flux density, is measured in tesla (in SI base units: kilogram per second² per ampere), which is equivalent to newton per meter per ampere. H and B differ in how they account for magnetization. In a vacuum, B and H are the same aside from units; but in a magnetized material, B/μ_0 and H differ by the magnetization M of the material at that point in the material.

Magnetic fields are produced by moving electric charges and the intrinsic magnetic moments of elementary particles associated with a fundamental quantum property, their spin. Magnetic fields and electric fields are interrelated, and are both components of the electromagnetic force, one of the four fundamental forces of nature.

Magnetic fields are widely used throughout modern technology, particularly in electrical engineering and electromechanics. Rotating magnetic fields are used in both electric motors and generators. The interaction of magnetic fields in electric devices such as transformers is studied in the discipline of magnetic circuits. Magnetic forces give information about the charge carriers in a material through the Hall effect. The Earth produces its own magnetic field, which shields the Earth's ozone layer from the solar wind and is important in navigation using a compass.

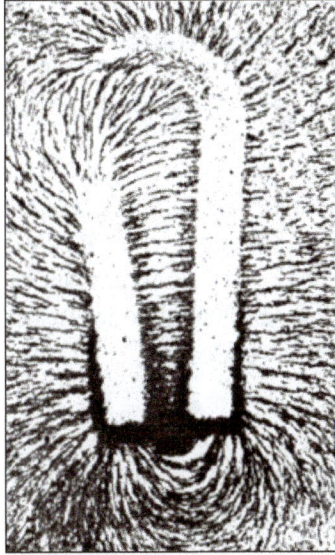

The shape of the magnetic field produced by a horseshoe magnet is revealed by the orientation of iron filings sprinkled on a piece of paper above the magnet.

Definitions, Units and Measurement

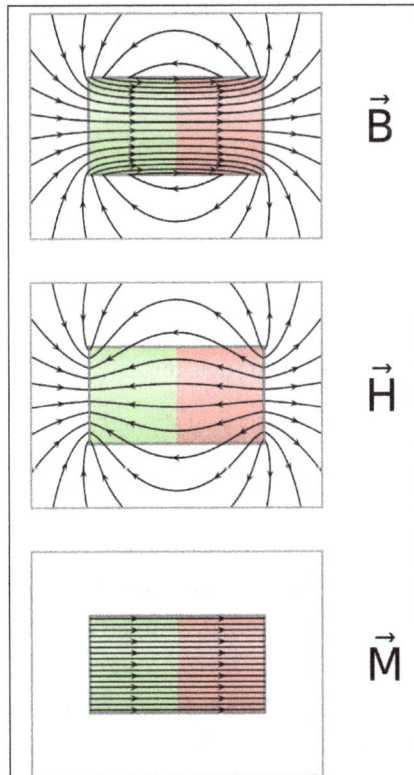

Comparison of B, H and M inside and outside a cylindrical bar magnet.

The B-field

Alternative names for B
•   Magnetic flux density
•   Magnetic induction
•   Magnetic field

The magnetic field can be defined in several equivalent ways based on the effects it has on its environment.

Often the magnetic field is defined by the force it exerts on a moving charged particle. Experiments in electrostatics show that a particle of charge q in an electric field E experiences a force $F = qE$. Other experiments show that a charged particle experiences a force proportional to its relative velocity to a current-carrying wire. The velocity dependent portion can be separated such that the force on the particle satisfies the *Lorentz force law*,

$$\mathbf{F} = q(\mathbf{E} + \mathbf{v} \times \mathbf{B}).$$

Here v is the particle's velocity and × denotes the cross product. The vector B is termed the magnetic field, and it is *defined* as the vector field necessary to make the Lorentz force law correctly describe the motion of a charged particle. This definition allows the determination of B in the following way:

> The command, "Measure the direction and magnitude of the vector B at such and such a place," calls for the following operations: Take a particle of known charge q. Measure the force on q at rest, to determine E. Then measure the force on the particle when its velocity is v; repeat with v in some other direction. Now find a B that makes the Lorentz force law fit all these results—that is the magnetic field at the place in question.

Alternatively, the magnetic field can be defined in terms of the torque it produces on a magnetic dipole.

The H-field

Alternative names for H
•   Magnetic field intensity
•   Magnetic field strength
•   Magnetic field
•   Magnetizing field

In addition to B, there is a quantity H, which is often called the *magnetic field*. In a vacuum, B and H are proportional to each other, with the multiplicative constant depending on the physical units. Inside a material they are different. The term "magnetic field" is historically reserved for H while using other terms for B. Informally, though, and formally for some recent textbooks mostly in physics, the term 'magnetic field' is used to describe B as well as or in place of H. There are many alternative names for both.

Units

In SI units, B is measured in teslas (symbol: T) and correspondingly Φ_B (magnetic flux) is measured in webers (symbol: Wb) so that a flux density of 1 Wb/m² is 1 tesla. The SI unit of tesla is equivalent to (newton·second)/(coulomb·metre). In Gaussian-cgs units, B is measured in gauss (symbol: G). (The conversion is 1 T = 10000 G.) One nanotesla is equivalent to 1 gamma (symbol: γ). The H-field is measured in amperes per metre (A/m) in SI units, and in oersteds (Oe) in cgs units.

Measurement

The finest precision for a magnetic field measurement was attained by Gravity Probe B at 5 aT (5×10^{-18} T).

Magnetometers are devices used to measure the local magnetic field. Important classes of magnetometers include using induction magnetometer (or search-coil magnetometer) which measure only varying magnetic fields, rotating coil magnetometer, Hall effect magnetometers, NMR magnetometers, SQUID magnetometers, and fluxgate magnetometers. The magnetic fields of distant astronomical objects are measured through their effects on local charged particles. For instance, electrons spiraling around a field line produce synchrotron radiation that is detectable in radio waves.

Magnetic Field Lines

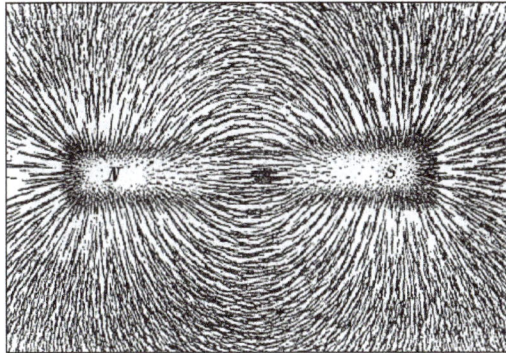

The direction of magnetic field lines represented by iron filings sprinkled on paper placed above a bar magnet.

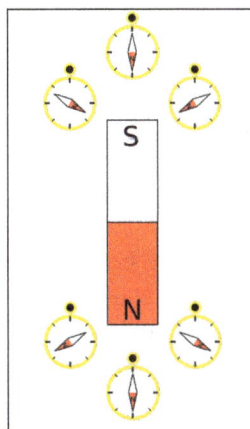

Compass needles point in the direction of the local magnetic field, towards
a magnet's south pole and away from its north pole.

Mapping the magnetic field of an object is simple in principle. First, measure the strength and direction of the magnetic field at a large number of locations (or at every point in space). Then, mark each location with an arrow (called a vector) pointing in the direction of the local magnetic field with its magnitude proportional to the strength of the magnetic field.

An alternative method to map the magnetic field is to "connect" the arrows to form magnetic *field lines*. The direction of the magnetic field at any point is parallel to the direction of nearby field lines, and the local density of field lines can be made proportional to its strength. Magnetic field lines are like streamlines in fluid flow, in that they represent something continuous, and a different resolution would show more or fewer lines.

An advantage of using magnetic field lines as a representation is that many laws of magnetism (and electromagnetism) can be stated completely and concisely using simple concepts such as the "number" of field lines through a surface. These concepts can be quickly "translated" to their mathematical form. For example, the number of field lines through a given surface is the surface integral of the magnetic field.

Various phenomena "display" magnetic field lines as though the field lines were physical phenomena. For example, iron filings placed in a magnetic field form lines that correspond to "field lines". Magnetic field "lines" are also visually displayed in polar auroras, in which plasma particle dipole interactions create visible streaks of light that line up with the local direction of Earth's magnetic field.

Field lines can be used as a qualitative tool to visualize magnetic forces. In ferromagnetic substances like iron and in plasmas, magnetic forces can be understood by imagining that the field lines exert a tension, (like a rubber band) along their length, and a pressure perpendicular to their length on neighboring field lines. "Unlike" poles of magnets attract because they are linked by many field lines; "like" poles repel because their field lines do not meet, but run parallel, pushing on each other. The rigorous form of this concept is the electromagnetic stress–energy tensor.

Magnetic Field and Permanent Magnets

Permanent magnets are objects that produce their own persistent magnetic fields. They are made of ferromagnetic materials, such as iron and nickel, that have been magnetized, and they have both a north and a south pole.

Magnetic Field of Permanent Magnets

The magnetic field of permanent magnets can be quite complicated, especially near the magnet. The magnetic field of a small straight magnet is proportional to the magnet's *strength* (called its magnetic dipole moment m). The equations are non-trivial and also depend on the distance from the magnet and the orientation of the magnet. For simple magnets, m points in the direction of a line drawn from the south to the north pole of the magnet. Flipping a bar magnet is equivalent to rotating its m by 180 degrees.

The magnetic field of larger magnets can be obtained by modeling them as a collection of a large number of small magnets called dipoles each having their own m. The magnetic field produced by the magnet then is the net magnetic field of these dipoles; any net force on the magnet is a result of adding up the forces on the individual dipoles.

There are two competing models for the nature of these dipoles. These two models produce two different magnetic fields, H and B. Outside a material, though, the two are identical (to a multiplicative constant) so that in many cases the distinction can be ignored. This is particularly true for magnetic fields, such as those due to electric currents, that are not generated by magnetic materials.

Magnetic Pole Model and the H-field

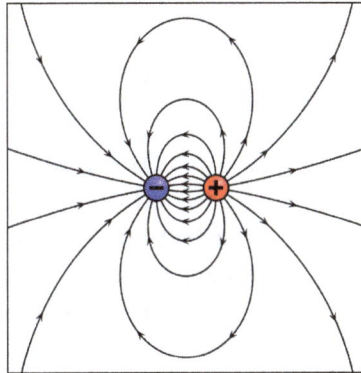

The magnetic pole model: two opposing poles, North (+) and South (−),
separated by a distance d produce a H-field (lines).

It is sometimes useful to model the force and torques between two magnets as due to magnetic poles repelling or attracting each other in the same manner as the Coulomb force between electric charges. In this model, a magnetic H-field is produced by *magnetic charges* that are "smeared" around each pole. These *magnetic charges* are in fact related to the magnetization field M.

The H-field, therefore, is analogous to the electric field E, which starts at a positive electric charge and ends at a negative electric charge. Near the north pole, therefore, all H-field lines point away from the north pole (whether inside the magnet or out) while near the south pole all H-field lines point toward the south pole (whether inside the magnet or out). Too, a north pole feels a force in the direction of the H-field while the force on the south pole is opposite to the H-field.

In the magnetic pole model, the elementary magnetic dipole m is formed by two opposite magnetic poles of pole strength q_m separated by a small distance vector d, such that m = q_md. The magnetic pole model predicts correctly the field H both inside and outside magnetic materials, in particular the fact that H is opposite to the magnetization field M inside a permanent magnet.

Since, it is based on the fictitious idea of a *magnetic charge density*, the pole model has limitations. Magnetic poles cannot exist apart from each other as electric charges can, but always come in north/south pairs. If a magnetized object is divided in half, a new pole appears on the surface of each piece, so each has a pair of complementary poles. The magnetic pole model does not account for magnetism that is produced by electric currents.

Amperian Loop Model and the B-field

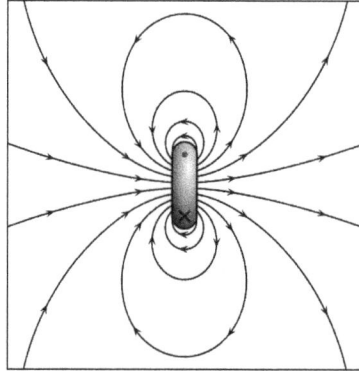

The Amperian loop model: A current loop (ring) that goes into the page at the x and comes out at the dot produces a B-field (lines). The north pole is to the right and the south to the left.

After Ørsted discovered that electric currents produce a magnetic field and Ampere discovered that electric currents attracted and repelled each other similar to magnets, it was natural to hypothesize that all magnetic fields are due to electric current loops. In this model developed by Ampere, the elementary magnetic dipole that makes up all magnets is a sufficiently small Amperian loop of current I. The dipole moment of this loop is $m = IA$ where A is the area of the loop.

These magnetic dipoles produce a magnetic B-field. One important property of the B-field produced this way is that magnetic B-field lines neither start nor end (mathematically, B is a solenoidal vector field); a field line either extends to infinity or wraps around to form a closed curve. To date, no exception to this rule has been found. Magnetic field lines exit a magnet near its north pole and enter near its south pole, but inside the magnet B-field lines continue through the magnet from the south pole back to the north. If a B-field line enters a magnet somewhere it has to leave somewhere else; it is not allowed to have an end point. Magnetic poles, therefore, always come in N and S pairs.

More formally, since all the magnetic field lines that enter any given region must also leave that region, subtracting the "number" of field lines that enter the region from the number that exit gives identically zero. Mathematically this is equivalent to:

$$\oint_S B \cdot dA = 0,$$

where the integral is a surface integral over the closed surface S (a closed surface is one that completely surrounds a region with no holes to let any field lines escape). Since dA points outward, the dot product in the integral is positive for B-field pointing out and negative for B-field pointing in.

There is also a corresponding differential form of this equation covered in Maxwell's equations below.

Force between Magnets

The force between two small magnets is quite complicated and depends on the strength and orientation of both magnets and the distance and direction of the magnets relative to each other. The force is particularly sensitive to rotations of the magnets due to magnetic torque. The force on each magnet depends on its magnetic moment and the magnetic field of the other.

To understand the force between magnets, it is useful to examine the *magnetic pole model* given above. In this model, the *H-field* of one magnet pushes and pulls on *both* poles of a second magnet. If this H-field is the same at both poles of the second magnet then there is no net force on that magnet since the force is opposite for opposite poles. If, however, the magnetic field of the first magnet is *nonuniform* (such as the H near one of its poles), each pole of the second magnet sees a different field and is subject to a different force. This difference in the two forces moves the magnet in the direction of increasing magnetic field and may also cause a net torque.

This is a specific example of a general rule that magnets are attracted (or repulsed depending on the orientation of the magnet) into regions of higher magnetic field. Any non-uniform magnetic field, whether caused by permanent magnets or electric currents, exerts a force on a small magnet in this way.

The details of the Amperian loop model are different and more complicated but yield the same result: that magnetic dipoles are attracted/repelled into regions of higher magnetic field. Mathematically, the force on a small magnet having a magnetic moment m due to a magnetic field B is:

$$F = \nabla\left(m \cdot B\right),$$

where the gradient ∇ is the change of the quantity m · B per unit distance and the direction is that of maximum increase of m · B. The dot product m · B = $mB\cos(\theta)$, where m and B represent the magnitude of the m and B vectors and θ is the angle between them. If m is in the same direction as B then the dot product is positive and the gradient points "uphill" pulling the magnet into regions of higher B-field (more strictly larger m · B). This equation is strictly only valid for magnets of zero size, but is often a good approximation for not too large magnets. The magnetic force on larger magnets is determined by dividing them into smaller regions each having their own m then summing up the forces on each of these very small regions.

Magnetic Torque on Permanent Magnets

If two like poles of two separate magnets are brought near each other, and one of the magnets is allowed to turn, it promptly rotates to align itself with the first. In this example, the magnetic field of the stationary magnet creates a *magnetic torque* on the magnet that is free to rotate. This magnetic torque τ tends to align a magnet's poles with the magnetic field lines. A compass, therefore, turns to align itself with Earth's magnetic field.

Magnetic torque is used to drive electric motors. In one simple motor design, a magnet is fixed to a freely rotating shaft and subjected to a magnetic field from an array of electromagnets. By continuously switching the electric current through each of the electromagnets, thereby flipping the polarity of their magnetic fields, like poles are kept next to the rotor; the resultant torque is transferred to the shaft.

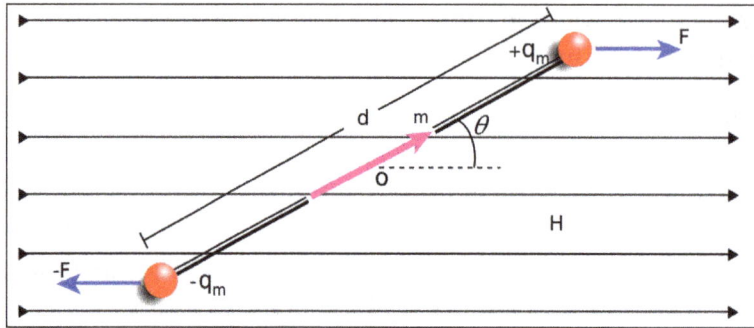

Torque on a Dipole: An H field (to right) causes equal but opposite forces on a
N pole (+q) and a S pole (−q) creating a torque.

As is the case for the force between magnets, the magnetic pole model leads more readily to the correct equation. Here, two equal and opposite magnetic charges experiencing the same H also experience equal and opposite forces. Since these equal and opposite forces are in different locations, this produces a torque proportional to the distance (perpendicular to the force) between them. With the definition of m as the pole strength times the distance between the poles, this leads to $\tau = \mu_0 mH\sin\theta$, where μ_0 is a constant called the vacuum permeability, measuring $4\pi\times10^{-7}$ V·s/(A·m) and θ is the angle between H and m.

The Amperian loop model also predicts the same magnetic torque. Here, it is the B field interacting with the Amperian current loop through a Lorentz force described below. Again, the results are the same although the models are completely different.

$$\tau = m \times B = \mu_0 m \times H,$$

where × represents the vector cross product. This equation includes all of the qualitative information included above. There is no torque on a magnet if m is in the same direction as the magnetic field. (The cross product is zero for two vectors that are in the same direction.) Further, all other orientations feel a torque that twists them toward the direction of magnetic field.

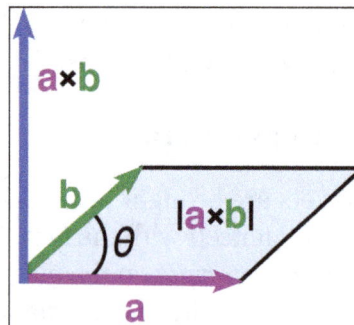

Cross product: $|a \times b| = a\,b\,\sin\theta$.

Mathematically, the torque τ on a small magnet is proportional both to the applied magnetic field and to the magnetic moment m of the magnet:

Magnetic Field and Electric Currents

Currents of electric charges both generate a magnetic field and feel a force due to magnetic B-fields.

Magnetic Field Due to Moving Charges and Electric Currents

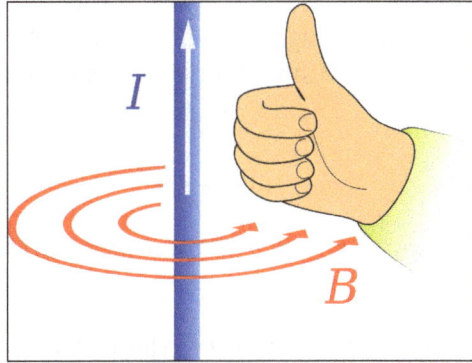

Right hand grip rule: a current flowing in the direction of the white arrow
produces a magnetic field shown by the red arrows.

All moving charged particles produce magnetic fields. Moving point charges, such as electrons, produce complicated but well known magnetic fields that depend on the charge, velocity, and acceleration of the particles.

Magnetic field lines form in concentric circles around a cylindrical current-carrying conductor, such as a length of wire. The direction of such a magnetic field can be determined by using the "right-hand grip rule". The strength of the magnetic field decreases with distance from the wire. (For an infinite length wire the strength is inversely proportional to the distance.)

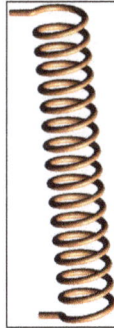

Solenoid

Bending a current-carrying wire into a loop concentrates the magnetic field inside the loop while weakening it outside. Bending a wire into multiple closely spaced loops to form a coil or "solenoid" enhances this effect. A device so formed around an iron core may act as an *electromagnet*, generating a strong, well-controlled magnetic field. An infinitely long cylindrical electromagnet has a uniform magnetic field inside, and no magnetic field outside. A finite length electromagnet produces a magnetic field that looks similar to that produced by a uniform permanent magnet, with its strength and polarity determined by the current flowing through the coil.

The magnetic field generated by a steady current I (a constant flow of electric charges, in which charge neither accumulates nor is depleted at any point) is described by the *Biot–Savart law*:

$$B = \frac{\mu_0 I}{4\pi} \int_{wire} \frac{d\ell \times \hat{r}}{r^2},$$

where the integral sums over the wire length where vector dℓ is the vector line element with direction in the same sense as the current I, μ_o is the magnetic constant, r is the distance between the location of dℓ and the location where the magnetic field is calculated, and $\hat{r}$ is a unit vector in the direction of r. For example, in the case of a sufficiently long, straight wire, this becomes:

$$|B| = \frac{\mu_0}{2\pi r} I$$

where $r = |\mathbf{r}|$. The direction is tangent to a circle perpendicular to the wire according to the right hand rule.

A slightly more general way of relating the current I to the B-field is through Ampère's law:

$$\oint B \cdot d\ell = \mu_0 I_{enc},$$

where the line integral is over any arbitrary loop and I_{enc} is the current enclosed by that loop. Ampère's law is always valid for steady currents and can be used to calculate the B-field for certain highly symmetric situations such as an infinite wire or an infinite solenoid.

In a modified form that accounts for time varying electric fields, Ampère's law is one of four Maxwell's equations that describe electricity and magnetism.

Force on Moving Charges and Current

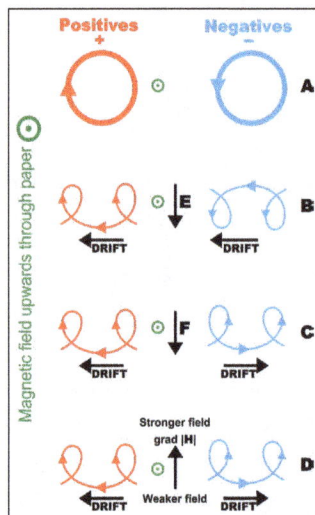

Charged particle drifts in a magnetic field with (A) no net force, (B) an electric field, E, (C) a charge independent force, F (e.g. gravity), and (D) in a homogeneous magnetic field, grad H.

Force on a Charged Particle

A charged particle moving in a B-field experiences a *sideways* force that is proportional to the strength of the magnetic field, the component of the velocity that is perpendicular to the magnetic field and the charge of the particle. This force is known as the *Lorentz force*, and is given by,

$$F = qE + qv \times B,$$

where F is the force, q is the electric charge of the particle, v is the instantaneous velocity of the particle, and B is the magnetic field (in teslas).

The Lorentz force is always perpendicular to both the velocity of the particle and the magnetic field that created it. When a charged particle moves in a static magnetic field, it traces a helical path in which the helix axis is parallel to the magnetic field, and in which the speed of the particle remains constant. Because the magnetic force is always perpendicular to the motion, the magnetic field can do no work on an isolated charge. It can only do work indirectly, via the electric field generated by a changing magnetic field. It is often claimed that the magnetic force can do work to a non-elementary magnetic dipole, or to charged particles whose motion is constrained by other forces, but this is incorrect because the work in those cases is performed by the electric forces of the charges deflected by the magnetic field.

Force on Current-carrying Wire

The force on a current carrying wire is similar to that of a moving charge as expected since a current carrying wire is a collection of moving charges. A current-carrying wire feels a force in the presence of a magnetic field. The Lorentz force on a macroscopic current is often referred to as the *Laplace force*. Consider a conductor of length ℓ, cross section A, and charge q due to electric current i. If this conductor is placed in a magnetic field of magnitude B that makes an angle θ with the velocity of charges in the conductor, the force exerted on a single charge q is:

$$F = qvB \sin \theta,$$

so, for N charges where,

$$N = n\ell A,$$

the force exerted on the conductor is,

$$f = FN = qvBn\ell A \sin \theta = Bi\ell \sin \theta,$$

where $i = nqvA$.

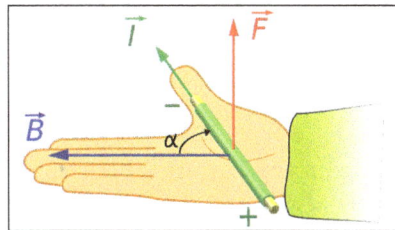

The right-hand rule: Pointing the thumb of the right hand in the direction of the conventional current, and the fingers in the direction of B, the force on the current points out of the palm. The force is reversed for a negative charge.

Direction of Force

The direction of force on a charge or a current can be determined by a mnemonic known as the *right-hand rule*. Using the right hand, pointing the thumb in the direction of the current, and the fingers in the direction of the magnetic field, the resulting force on the charge points outwards from the palm. The force on a negatively charged particle is in the opposite direction. If both the speed and the charge are reversed then the direction of the force remains the same. For that reason a magnetic field measurement (by itself) cannot distinguish whether there is a positive charge moving to the right or a negative charge moving to the left. (Both of these cases produce the same current.) On the other hand, a magnetic field combined with an electric field *can* distinguish between these. An alternative mnemonic to the right hand rule is Fleming's left-hand rule.

Relation between H and B

The formulas derived for the magnetic field above are correct when dealing with the entire current. A magnetic material placed inside a magnetic field, though, generates its own bound current, which can be a challenge to calculate. (This bound current is due to the sum of atomic sized current loops and the spin of the subatomic particles such as electrons that make up the material.) The H-field as defined above helps factor out this bound current; but to see how, it helps to introduce the concept of *magnetization* first.

Magnetization

The *magnetization* vector field M represents how strongly a region of material is magnetized. It is defined as the net magnetic dipole moment per unit volume of that region. The magnetization of a uniform magnet is therefore a material constant, equal to the magnetic moment m of the magnet divided by its volume. Since the SI unit of magnetic moment is A·m², the SI unit of magnetization M is ampere per meter, identical to that of the H-field.

The magnetization M field of a region points in the direction of the average magnetic dipole moment in that region. Magnetization field lines, therefore, begin near the magnetic south pole and ends near the magnetic north pole. (Magnetization does not exist outside the magnet.)

In the Amperian loop model, the magnetization is due to combining many tiny Amperian loops to form a resultant current called *bound current*. This bound current, then, is the source of the magnetic B field due to the magnet. Given the definition of the magnetic dipole, the magnetization field follows a similar law to that of Ampere's law:

$$\oint M \cdot d\ell = I_b,$$

where the integral is a line integral over any closed loop and I_b is the bound current enclosed by that closed loop.

In the magnetic pole model, magnetization begins at and ends at magnetic poles. If a given region, therefore, has a net positive "magnetic pole strength" (corresponding to a north pole) then it has more magnetization field lines entering it than leaving it. Mathematically this is equivalent to:

$$\oint_S \mu_0 M \cdot dA = -q_M,$$

where the integral is a closed surface integral over the closed surface S and q_M is the "magnetic charge" (in units of magnetic flux) enclosed by S. (A closed surface completely surrounds a region with no holes to let any field lines escape.) The negative sign occurs because the magnetization field moves from south to north.

H-field and Magnetic Materials

In SI units, the H-field is related to the B-field by,

$$H \equiv \frac{B}{\mu_0} - M.$$

In terms of the H-field, Ampere's law is,

$$\oint H \cdot d\ell = \oint \left(\frac{B}{\mu_0} - M \right) \cdot d\ell = I_{tot} - I_b = I_f,$$

where I_f represents the 'free current' enclosed by the loop so that the line integral of H does not depend at all on the bound currents.

Ampere's law leads to the boundary condition,

$$\left(H_1^{\parallel} - H_2^{\parallel} \right) = K_f \times \hat{n},$$

where K_f is the surface free current density and the unit normal $\hat{n}$ points in the direction from medium 2 to medium 1.

Similarly, a surface integral of H over any closed surface is independent of the free currents and picks out the "magnetic charges" within that closed surface:

$$\oint_S \mu_0 H \cdot dA = \oint_S (B - \mu_0 M) \cdot dA = 0 - (-q_M) = q_M,$$

which does not depend on the free currents.

The H-field, therefore, can be separated into two independent parts:

$$H = H_0 + H_d,$$

where H_0 is the applied magnetic field due only to the free currents and H_d is the demagnetizing field due only to the bound currents.

The magnetic H-field, therefore, re-factors the bound current in terms of "magnetic charges". The H field lines loop only around "free current" and, unlike the magnetic B field, begins and ends near magnetic poles as well.

Magnetism

Most materials respond to an applied B-field by producing their own magnetization M and therefore their own B-fields. Typically, the response is weak and exists only when the magnetic field is applied. The term *magnetism* describes how materials respond on the microscopic level to an applied magnetic field and is used to categorize the magnetic phase of a material. Materials are divided into groups based upon their magnetic behavior:

- Diamagnetic materials produce a magnetization that opposes the magnetic field.

- Paramagnetic materials produce a magnetization in the same direction as the applied magnetic field.

- Ferromagnetic materials and the closely related ferrimagnetic materials and antiferromagnetic materials can have a magnetization independent of an applied B-field with a complex relationship between the two fields.

- Superconductors (and ferromagnetic superconductors) are materials that are characterized by perfect conductivity below a critical temperature and magnetic field. They also are highly magnetic and can be perfect diamagnets below a lower critical magnetic field. Superconductors often have a broad range of temperatures and magnetic fields (the so-named mixed state) under which they exhibit a complex hysteretic dependence of M on B.

In the case of paramagnetism and diamagnetism, the magnetization M is often proportional to the applied magnetic field such that:

$$B = \mu H,$$

where μ is a material dependent parameter called the permeability. In some cases the permeability may be a second rank tensor so that H may not point in the same direction as B. These relations between B and H are examples of constitutive equations. However, superconductors and ferromagnets have a more complex B-to-H relation.

Energy Stored in Magnetic Fields

Energy is needed to generate a magnetic field both to work against the electric field that a changing magnetic field creates and to change the magnetization of any material within the magnetic field. For non-dispersive materials, this same energy is released when the magnetic field is destroyed so that the energy can be modeled as being stored in the magnetic field.

For linear, non-dispersive, materials (such that B = μH where μ is frequency-independent), the energy density is:

$$u = \frac{B \cdot H}{2} = \frac{B \cdot B}{2\mu} = \frac{\mu H \cdot H}{2}.$$

If there are no magnetic materials around then μ can be replaced by μ_0. The above equation cannot be used for nonlinear materials, though; a more general expression given below must be used.

In general, the incremental amount of work per unit volume δW needed to cause a small change of magnetic field δB is:

$$\delta W = H \cdot \delta B.$$

Once the relationship between H and B is known this equation is used to determine the work needed to reach a given magnetic state. For hysteretic materials such as ferromagnets and superconductors, the work needed also depends on how the magnetic field is created. For linear non-dispersive materials, though, the general equation leads directly to the simpler energy density equation given above.

Important uses and Examples of Magnetic Field

Earth's Magnetic Field

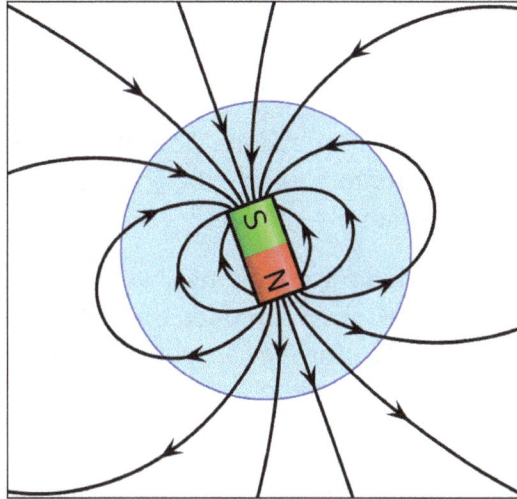

A sketch of Earth's magnetic field representing the source of the field as a magnet.
The geographic north pole is above the south pole of the magnet.

The Earth's magnetic field is produced by convection of a liquid iron alloy in the outer core. In a dynamo process, the movements drive a feedback process in which electric currents create electric and magnetic fields that in turn act on the currents.

The field at the surface of the Earth is approximately the same as if a giant bar magnet were positioned at the center of the Earth and tilted at an angle of about 11° off the rotational axis of the Earth. The north pole of a magnetic compass needle points roughly north, toward the North Magnetic Pole. However, because a magnetic pole is attracted to its opposite, the North Magnetic Pole is actually the south pole of the geomagnetic field. This confusion in terminology arises because the pole of a magnet is defined by the geographical direction it points.

Earth's magnetic field is not constant—the strength of the field and the location of its poles vary. Moreover, the poles periodically reverse their orientation in a process called geomagnetic reversal. The most recent reversal occurred 780,000 years ago.

Record-high Fields

The largest magnetic field produced over a macroscopic volume is 2.8 kT. The largest magnetic fields produced in a laboratory occur in particle accelerators, such as RHIC, inside the collisions of heavy ions, where fields reach 10^{14} T. Magnetars have the strongest known magnetic fields of any naturally occurring object, ranging from 0.1 to 100 GT (10^8 to 10^{11} T).

Rotating Magnetic Fields

The *rotating magnetic field* is a key principle in the operation of alternating-current motors. A permanent magnet in such a field rotates so as to maintain its alignment with the external field. This effect was conceptualized by Nikola Tesla, and later utilized in his and others' early AC (alternating current) electric motors.

A rotating magnetic field can be constructed using two orthogonal coils with 90 degrees phase difference in their AC currents. However, in practice such a system would be supplied through a three-wire arrangement with unequal currents.

This inequality would cause serious problems in standardization of the conductor size and so, to overcome it, three-phase systems are used where the three currents are equal in magnitude and have 120 degrees phase difference. Three similar coils having mutual geometrical angles of 120 degrees create the rotating magnetic field in this case. The ability of the three-phase system to create a rotating field, utilized in electric motors, is one of the main reasons why three-phase systems dominate the world's electrical power supply systems.

Synchronous motors use DC-voltage-fed rotor windings, which lets the excitation of the machine be controlled—and induction motors use short-circuited rotors (instead of a magnet) following the rotating magnetic field of a multicoiled stator. The short-circuited turns of the rotor develop eddy currents in the rotating field of the stator, and these currents in turn move the rotor by the Lorentz force.

In 1882, Nikola Tesla identified the concept of the rotating magnetic field. In 1885, Galileo Ferraris independently researched the concept. In 1888, Tesla gained U.S. Patent 381,968 for his work. Also in 1888, Ferraris published his research in a paper to the *Royal Academy of Sciences* in Turin.

Hall Effect

The charge carriers of a current-carrying conductor placed in a transverse magnetic field experience a sideways Lorentz force; this results in a charge separation in a direction perpendicular to the current and to the magnetic field. The resultant voltage in that direction is proportional to the applied magnetic field. This is known as the *Hall effect*.

The *Hall effect* is often used to measure the magnitude of a magnetic field. It is used as well to find the sign of the dominant charge carriers in materials such as semiconductors (negative electrons or positive holes).

Magnetic Circuits

An important use of H is in *magnetic circuits* where $B = \mu H$ inside a linear material. Here, μ is the magnetic permeability of the material. This result is similar in form to Ohm's law $J = \sigma E$, where J is the current density, σ is the conductance and E is the electric field. Extending this analogy, the counterpart to the macroscopic Ohm's law ($I = V/R$) is:

$$\Phi = \frac{F}{R_m},$$

where $\Phi = \int B \cdot dA$ is the magnetic flux in the circuit, $F = \int H \cdot d\ell$ is the magnetomotive force applied to the circuit, and R_{m} is the reluctance of the circuit. Here the reluctance R_{m} is a quantity similar in nature to resistance for the flux.

Using this analogy it is straightforward to calculate the magnetic flux of complicated magnetic field geometries, by using all the available techniques of circuit theory.

Magnetic Field Shape Descriptions

Schematic quadrupole magnet ("*four-pole*") magnetic field. There are four steel pole tips, two opposing magnetic north poles and two opposing magnetic south poles.

- An azimuthal magnetic field is one that runs east–west.

- A meridional magnetic field is one that runs north–south. In the solar dynamo model of the Sun, differential rotation of the solar plasma causes the meridional magnetic field to stretch into an azimuthal magnetic field, a process called the omega-effect. The reverse process is called the alpha-effect.

- A dipole magnetic field is one seen around a bar magnet or around a charged elementary particle with nonzero spin.

- A quadrupole magnetic field is one seen, for example, between the poles of four bar magnets. The field strength grows linearly with the radial distance from its longitudinal axis.

- A solenoidal magnetic field is similar to a dipole magnetic field, except that a solid bar magnet is replaced by a hollow electromagnetic coil magnet.

- A toroidal magnetic field occurs in a doughnut-shaped coil, the electric current spiraling around the tube-like surface, and is found, for example, in a tokamak.

- A poloidal magnetic field is generated by a current flowing in a ring, and is found, for example, in a tokamak.

- A radial magnetic field is one in which field lines are directed from the center outwards, similar to the spokes in a bicycle wheel. An example can be found in a loudspeaker transducers (driver).

- A helical magnetic field is corkscrew-shaped, and sometimes seen in space plasmas such as the Orion Molecular Cloud.

Magnetic Dipoles

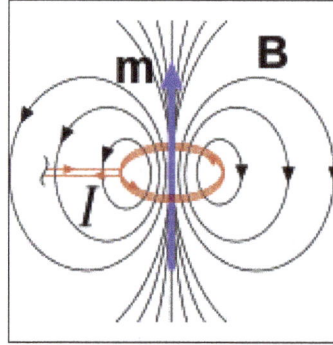

Magnetic field lines around a "magnetostatic dipole" pointing to the right.

The magnetic field of a magnetic dipole is depicted in the figure. From outside, the ideal magnetic dipole is identical to that of an ideal electric dipole of the same strength. Unlike the electric dipole, a magnetic dipole is properly modeled as a current loop having a current I and an area a. Such a current loop has a magnetic moment of:

$$m = Ia,$$

where the direction of m is perpendicular to the area of the loop and depends on the direction of the current using the right-hand rule. An ideal magnetic dipole is modeled as a real magnetic dipole whose area a has been reduced to zero and its current I increased to infinity such that the product $m = Ia$ is finite. This model clarifies the connection between angular momentum and magnetic moment, which is the basis of the Einstein–de Haas effect *rotation by magnetization* and its inverse, the Barnett effect or *magnetization by rotation*. Rotating the loop faster (in the same direction) increases the current and therefore the magnetic moment, for example.

It is sometimes useful to model the magnetic dipole similar to the electric dipole with two equal but opposite magnetic charges (one south the other north) separated by distance d. This model produces an H-field not a B-field. Such a model is deficient, though, both in that there are no magnetic charges and in that it obscures the link between electricity and magnetism. Further, it fails to explain the inherent connection between angular momentum and magnetism.

Magnetic Monopole (Hypothetical)

A *magnetic monopole* is a hypothetical particle (or class of particles) that has, as its name suggests, only one magnetic pole (either a north pole or a south pole). In other words, it would possess a "magnetic charge" analogous to an electric charge. Magnetic field lines would start or end on magnetic monopoles, so if they exist, they would give exceptions to the rule that magnetic field lines neither start nor end.

Modern interest in this concept stems from particle theories, notably Grand Unified Theories and superstring theories, that predict either the existence, or the possibility, of magnetic monopoles. These theories and others have inspired extensive efforts to search for monopoles. Despite these efforts, no magnetic monopole has been observed to date.

In recent research, materials known as spin ices can simulate monopoles, but do not contain actual monopoles.

Gauss's Law for Magnetism

In physics, Gauss's law for magnetism is one of the four Maxwell's equations that underlie classical electrodynamics. It states that the magnetic field B has divergence equal to zero, in other words, that it is a solenoidal vector field. It is equivalent to the statement that magnetic monopoles do not exist. Rather than "magnetic charges", the basic entity for magnetism is the magnetic dipole. (If monopoles were ever found, the law would have to be modified.)

Gauss's law for magnetism can be written in two forms, a *differential form* and an *integral form*. These forms are equivalent due to the divergence theorem.

The name "Gauss's law for magnetism" is not universally used. The law is also called "Absence of free magnetic poles"; one reference even explicitly says the law has "no name". It is also referred to as the "transversality requirement" because for plane waves it requires that the polarization be transverse to the direction of propagation.

Differential Form

The differential form for Gauss's law for magnetism is:

$$\nabla \cdot B = 0$$

where $\nabla \cdot$ denotes divergence, and B is the magnetic field.

Integral Form

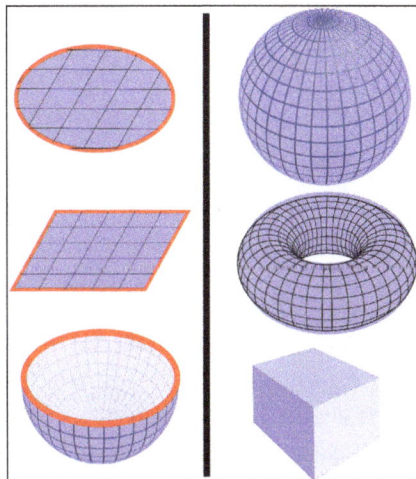

Definition of a closed surface.

Left: Some examples of closed surfaces include the surface of a sphere, surface of a torus, and surface of a cube. The magnetic flux through any of these surfaces is zero.

Right: Some examples of non-closed surfaces include the disk surface, square surface, or hemisphere surface. They all have boundaries (red lines) and they do not fully enclose a 3D volume. The magnetic flux through these surfaces is *not necessarily zero*.

The integral form of Gauss's law for magnetism states:

$$\oiint_S B \cdot dA = 0$$

where S is any closed surface, and dA is a vector, whose magnitude is the area of an infinitesimal piece of the surface S, and whose direction is the outward-pointing surface normal.

The left-hand side of this equation is called the net flux of the magnetic field out of the surface, and Gauss's law for magnetism states that it is always zero.

The integral and differential forms of Gauss's law for magnetism are mathematically equivalent, due to the divergence theorem. That said, one or the other might be more convenient to use in a particular computation.

The law in this form states that for each volume element in space, there are exactly the same number of "magnetic field lines" entering and exiting the volume. No total "magnetic charge" can build up in any point in space. For example, the south pole of the magnet is exactly as strong as the north pole, and free-floating south poles without accompanying north poles (magnetic monopoles) are not allowed. In contrast, this is not true for other fields such as electric fields or gravitational fields, where total electric charge or mass can build up in a volume of space.

Vector Potential

Due to the Helmholtz decomposition theorem, Gauss's law for magnetism is equivalent to the following statement:

There exists a vector field A such that,

$$B = \nabla \times A.$$

The vector field A is called the magnetic vector potential.

Note that there is more than one possible A which satisfies this equation for a given B field. In fact, there are infinitely many: any field of the form $\nabla\phi$ can be added onto A to get an alternative choice for A, by the identity:

$$\nabla \times A = \nabla \times (A + \nabla\phi)$$

since the curl of a gradient is the zero vector field:

$$\nabla \times \nabla\phi = \mathbf{0}$$

This arbitrariness in A is called gauge freedom.

Field Lines

The magnetic field B, like any vector field, can be depicted via field lines (also called *flux lines*) – that is, a set of curves whose direction corresponds to the direction of B, and whose areal density is proportional to the magnitude of B. Gauss's law for magnetism is equivalent to the statement that the field lines have neither a beginning nor an end: Each one either forms a closed loop, winds around forever without ever quite joining back up to itself exactly, or extends to infinity.

Modification if Magnetic Monopoles Exist

If magnetic monopoles were discovered, then Gauss's law for magnetism would state the divergence of B would be proportional to the *magnetic charge density* ρ_m, analogous to Gauss's law for electric field. For zero net magnetic charge density ($\rho_m = 0$), the original form of Gauss's magnetism law is the result.

The modified formula in SI units is not standard; in one variation, magnetic charge has units of webers, in another it has units of ampere-meters.

Units	Equation
cgs units	$\nabla \cdot B = 4\pi\rho_m$
SI units (weber convention)	$\nabla \cdot B = \rho_m$
SI units (ampere-meter convention)	$\nabla \cdot B = \mu_0\rho_m$

where μ_0 is the vacuum permeability.

So far, no magnetic monopoles have been found, despite extensive search.

Ampère's Circuital Law

In classical electromagnetism, Ampère's circuital law (not to be confused with Ampère's force law that André-Marie Ampère discovered in 1823) relates the integrated magnetic field around a closed loop to the electric current passing through the loop. James Clerk Maxwell (not Ampère) derived it using hydrodynamics in his 1861 paper "On Physical Lines of Force" and it is now one of the Maxwell equations, which form the basis of classical electromagnetism.

Maxwell's Original Circuital Law

The original form of Maxwell's circuital law, which he derived in his 1855 paper "On Faraday's Lines of Force" based on an analogy to hydrodynamics, relates magnetic fields to electric currents that produce them. It determines the magnetic field associated with a given current, or the current associated with a given magnetic field.

The original circuital law is only a correct law of physics in a magnetostatic situation, where the system is static except possibly for continuous steady currents within closed loops. For systems with electric fields that change over time, the original law must be modified to include a term known as Maxwell's correction.

Equivalent Forms

The original circuital law can be written in several different forms, which are all ultimately equivalent:

- An "integral form" and a "differential form". The forms are exactly equivalent, and related by the Kelvin–Stokes theorem.

- Forms using SI units, and those using cgs units. Other units are possible, but rare.

- Forms using either B or H magnetic fields. These two forms use the total current density and free current density, respectively. The B and H fields are related by the constitutive equation: $B = \mu_0 H$ where μ_0 is the magnetic constant.

Explanation

The integral form of the original circuital law is a line integral of the magnetic field around some closed curve C (arbitrary but must be closed). The curve C in turn bounds both a surface S which the electric current passes through (again arbitrary but not closed—since no three-dimensional volume is enclosed by S), and encloses the current. The mathematical statement of the law is a relation between the total amount of magnetic field around some path (line integral) due to the current which passes through that enclosed path (surface integral).

In terms of total current, (which is the sum of both free current and bound current) the line integral of the magnetic B-field (in teslas, T) around closed curve C is proportional to the total current I_{enc} passing through a surface S (enclosed by C). In terms of free current, the line integral of the magnetic H-field (in amperes per metre, A·m^{-1}) around closed curve C equals the free current $I_{f,enc}$ through a surface S.

Forms of the Original Circuital Law Written in SI Units

	Integral form	Differential form
Using B-field and total current	$\oint_C B \cdot dl = \mu_0 \iint_S J \cdot dS = \mu_0 I_{enc}$	$\nabla \times B = \mu_0 J$
Using H-field and free current	$\oint_C H \cdot dl = \iint_S J_f \cdot dS = I_{f,enc}$	$\nabla \times H = J_f$

- J is the total current density (in amperes per square metre, A·m^{-2}),

- J_f is the free current density only,

- $\oint_C$ is the closed line integral around the closed curve C,

- $\iint_S$ denotes a 2-D surface integral over S enclosed by C,

- · is the vector dot product,

- dl is an infinitesimal element (a differential) of the curve C (i.e. a vector with magnitude equal to the length of the infinitesimal line element, and direction given by the tangent to the curve C),

- dS is the vector area of an infinitesimal element of surface S (that is, a vector with magnitude equal to the area of the infinitesimal surface element, and direction normal to surface S. The direction of the normal must correspond with the orientation of C by the right hand rule,

- $\nabla \times$ is the curl operator.

Ambiguities and Sign Conventions

There are a number of ambiguities in the above definitions that require clarification and a choice of convention.

- First, three of these terms are associated with sign ambiguities: the line integral $\oint_C$ could go around the loop in either direction (clockwise or counterclockwise); the vector area dS could point in either of the two directions normal to the surface; and I_{enc} is the net current passing through the surface S, meaning the current passing through in one direction, minus the current in the other direction—but either direction could be chosen as positive. These ambiguities are resolved by the right-hand rule: With the palm of the right-hand toward the area of integration, and the index-finger pointing along the direction of line-integration, the outstretched thumb points in the direction that must be chosen for the vector area dS. Also the current passing in the same direction as dS must be counted as positive. The right hand grip rule can also be used to determine the signs.

- Second, there are infinitely many possible surfaces S that have the curve C as their border. (Imagine a soap film on a wire loop, which can be deformed by moving the wire). Which of those surfaces is to be chosen? If the loop does not lie in a single plane, for example, there is no one obvious choice. The answer is that it does not matter; by Stoke's theorem, the integral is the same for any surface with boundary C, since the integrand is the curl of a smooth field (i.e. exact). In practice, one usually chooses the most convenient surface (with the given boundary) to integrate over.

Free Current versus Bound Current

The electric current that arises in the simplest textbook situations would be classified as "free current"—for example, the current that passes through a wire or battery. In contrast, "bound current" arises in the context of bulk materials that can be magnetized and/or polarized. (All materials can to some extent.)

When a material is magnetized (for example, by placing it in an external magnetic field), the electrons remain bound to their respective atoms, but behave as if they were orbiting the nucleus in a particular direction, creating a microscopic current. When the currents from all these atoms are put together, they create the same effect as a macroscopic current, circulating perpetually around the magnetized object. This magnetization current J_M is one contribution to "bound current".

The other source of bound current is bound charge. When an electric field is applied, the positive and negative bound charges can separate over atomic distances in polarizable materials, and when the bound charges move, the polarization changes, creating another contribution to the "bound current", the polarization current J_p.

The total current density J due to free and bound charges is then:

$$J = J_f + J_M + J_P,$$

with J_f the "free" or "conduction" current density.

All current is fundamentally the same, microscopically. Nevertheless, there are often practical reasons for wanting to treat bound current differently from free current. For example, the bound current usually originates over atomic dimensions, and one may wish to take advantage of a simpler theory intended for larger dimensions. The result is that the more microscopic Ampère's circuital law, expressed in terms of B and the microscopic current (which includes free, magnetization and polarization currents), is sometimes put into the equivalent form below in terms of H and the free current only.

Shortcomings of the Original Formulation of the Circuital Law

There are two important issues regarding the circuital law that require closer scrutiny. First, there is an issue regarding the continuity equation for electrical charge. In vector calculus, the identity for the divergence of a curl states that the divergence of the curl of a vector field must always be zero. Hence,

$$\nabla \cdot (\nabla \times B) = 0,$$

and so the original Ampère's circuital law implies that,

$$\nabla \cdot J = 0.$$

But in general, reality follows the continuity equation for electric charge:

$$\nabla \cdot J = -\frac{\partial \rho}{\partial t},$$

which is nonzero for a time-varying charge density. An example occurs in a capacitor circuit where time-varying charge densities exist on the plates.

Second, there is an issue regarding the propagation of electromagnetic waves. For example, in free space, where,

$$J = 0.$$

The circuital law implies that,

$$\nabla \times B = 0,$$

but to maintain consistency with the continuity equation for electric charge, we must have,

$$\nabla \times B = \frac{1}{c^2} \frac{\partial E}{\partial t}.$$

To treat these situations, the contribution of displacement current must be added to the current term in the circuital law.

James Clerk Maxwell conceived of displacement current as a polarization current in the dielectric vortex sea, which he used to model the magnetic field hydrodynamically and mechanically. He added this displacement current to Ampère's circuital law at equation 112 in his 1861 paper "On Physical Lines of Force".

Displacement Current

In free space, the displacement current is related to the time rate of change of electric field. In a dielectric the above contribution to displacement current is present too, but a major contribution to the displacement current is related to the polarization of the individual molecules of the dielectric material. Even though charges cannot flow freely in a dielectric, the charges in molecules can move a little under the influence of an electric field. The positive and negative charges in molecules separate under the applied field, causing an increase in the state of polarization, expressed as the polarization density P. A changing state of polarization is equivalent to a current.

Both contributions to the displacement current are combined by defining the displacement current as:

$$J_D = \frac{\partial}{\partial t} D(r,t),$$

where the electric displacement field is defined as:

$$D = \varepsilon_0 E + P = \varepsilon_0 \varepsilon_r E,$$

where ε_0 is the electric constant, ε_r the relative static permittivity, and P is the polarization density. Substituting this form for D in the expression for displacement current, it has two components:

$$J_D = \varepsilon_0 \frac{\partial E}{\partial t} + \frac{\partial P}{\partial t}.$$

The first term on the right hand side is present everywhere, even in a vacuum. It doesn't involve any actual movement of charge, but it nevertheless has an associated magnetic field, as if it were an actual current.

The second term on the right hand side is the displacement current as originally conceived by Maxwell, associated with the polarization of the individual molecules of the dielectric material.

Maxwell's original explanation for displacement current focused upon the situation that occurs in dielectric media. In the modern post-aether era, the concept has been extended to apply to situations with no material media present, for example, to the vacuum between the plates of a charging vacuum capacitor. The displacement current is justified today because it serves several requirements of an electromagnetic theory: correct prediction of magnetic fields in regions where no free current flows; prediction of wave propagation of electromagnetic fields; and conservation of electric charge in cases where charge density is time-varying.

Extending the Original Law: The Maxwell–Ampère Equation

Next, the circuital equation is extended by including the polarization current, thereby remedying the limited applicability of the original circuital law.

Treating free charges separately from bound charges, The equation including Maxwell's correction in terms of the H-field is (the H-field is used because it includes the magnetization currents, so J_M does not appear explicitly:

$$\oint_C H \cdot dl = \iint_S \left(J_f + \frac{\partial D}{\partial t} \right) \cdot dS$$

(integral form), where H is the magnetic H field (also called "auxiliary magnetic field", "magnetic field intensity", or just "magnetic field"), D is the electric displacement field, and J_f is the enclosed conduction current or free current density. In differential form,

$$\nabla \times H = J_f + \frac{\partial D}{\partial t}.$$

On the other hand, treating all charges on the same footing (disregarding whether they are bound or free charges), the generalized Ampère's equation, also called the Maxwell–Ampère equation, is in integral form:

$$\oint_C B \cdot dl = \iint_S \left(\mu_0 J + \mu_0 \varepsilon_0 \frac{\partial E}{\partial t} \right) \cdot dS$$

In differential form,

$$\nabla \times B = \mu_0 J + \mu_0 \varepsilon_0 \frac{\partial E}{\partial t}$$

In both forms J includes magnetization current density as well as conduction and polarization current densities. That is, the current density on the right side of the Ampère–Maxwell equation is:

$$J_f + J_D + J_M = J_f + J_P + J_M + \varepsilon_0 \frac{\partial E}{\partial t} = J + \varepsilon_0 \frac{\partial E}{\partial t},$$

where current density J_D is the *displacement current*, and J is the current density contribution actually due to movement of charges, both free and bound. Because $\nabla \cdot D = \rho$, the charge continuity issue with Ampère's original formulation is no longer a problem. Because of the term in $\varepsilon_0 \partial E/\partial t$, wave propagation in free space now is possible.

With the addition of the displacement current, Maxwell was able to hypothesize (correctly) that light was a form of electromagnetic wave.

Proof of Equivalence

Proof that the formulations of the circuital law in terms of free current are equivalent to the formulations involving total current.

In this proof, we will show that the equation,

$$\nabla \times H = J_f + \frac{\partial D}{\partial t}$$

is equivalent to the equation,

$$\frac{1}{\mu_0}(\nabla \times B) = J + \varepsilon_0 \frac{\partial E}{\partial t}.$$

Note that we are only dealing with the differential forms, not the integral forms, but that is sufficient since the differential and integral forms are equivalent in each case, by the Kelvin–Stokes theorem.

We introduce the polarization density P, which has the following relation to E and D:

$$D = \varepsilon_0 E + P.$$

Next, we introduce the magnetization density M, which has the following relation to B and H:

$$\frac{1}{\mu_0} B = H + M$$

and the following relation to the bound current:

$$J_{bound} = \nabla \times M + \frac{\partial P}{\partial t}$$

$$= J_M + J_P,$$

where,

$$J_M = \nabla \times M,$$

is called the magnetization current density, and

$$J_P = \frac{\partial P}{\partial t},$$

is the polarization current density. Taking the equation for B:

$$\frac{1}{\mu_0}(\nabla \times B) = \nabla \times (H + M)$$

$$= \nabla \times H + J_M$$

$$= J_f + J_P + \varepsilon_0 \frac{\partial E}{\partial t} + J_M.$$

Consequently, referring to the definition of the bound current:

$$\frac{1}{\mu_0}(\nabla \times B) = J_f + J_{bound} + \varepsilon_0 \frac{\partial E}{\partial t}$$

$$= J + \varepsilon_0 \frac{\partial E}{\partial t},$$

as was to be shown.

Ampère's Circuital Law in CGS Units

In cgs units, the integral form of the equation, including Maxwell's correction, reads,

$$\oint_C B \cdot dl = \frac{1}{c}\iint_S \left(4\pi J + \frac{\partial E}{\partial t}\right) \cdot dS,$$

where c is the speed of light.

The differential form of the equation (again, including Maxwell's correction) is,

$$\nabla \times B = \frac{1}{c}\left(4\pi J + \frac{\partial E}{\partial t}\right).$$

Magnetic Dipole Moment

A magnetic moment is a quantity that represents the magnetic strength and orientation of a magnet or other object that produces a magnetic field. More precisely, a magnetic moment refers to a magnetic dipole moment, the component of magnetic moment that can be represented by a magnetic dipole. A magnetic dipole is a magnetic north pole and a magnetic south pole separated by a small distance.

Magnetic dipole moments have dimensions of current times area or energy divided by magnetic flux density. The unit for dipole moment in metre–kilogram– second–ampere is ampere-square metre. The unit in centimetre–gram–second, electromagnetic system, is the erg (unit of energy) per gauss (unit of magnetic flux density). One thousand ergs per gauss equal one ampere-square metre.

Derivation of Magnetic Dipole Moment Formula

Magnetic Dipole moment- The magnetic field, B due to a current loop carrying a current i of radius, R at a distance l along its axis is given by:

$$B = \frac{\mu_0 i R^2}{2(R^2 + l^2)^{\frac{3}{2}}}$$

Now if we consider a point very far from the current loop such that l>>R, then we can approximate the field as:

$$B = \frac{\mu_0 i R^2}{2l^3 \left(\left(\frac{R}{l} \right)^2 + 1 \right)^{\frac{3}{2}}} \approx \frac{\mu_0 i R^2}{2l^3} \equiv \frac{\mu_0}{4\pi} \frac{2i \left(\pi R^2 \right)}{l^3}$$

Now, the area of the loop, A is,

$$A = \pi R^2$$

Thus, the magnetic field can be written as,

$$B = \frac{\mu_0}{4\pi} \frac{2iA}{l^3} = \frac{\mu_0}{4\pi} \frac{2\mu}{l^3}$$

We can write this new quantity μ as a vector that points along the magnetic field, so that,

$$\vec{B} = \frac{\mu_0}{4\pi} \frac{2\vec{\mu}}{l^3}$$

Notice the astounding similarity to the electric dipole field:

$$\vec{E} = \frac{1}{4\pi\varepsilon_0} \frac{2\vec{p}}{r^3}$$

hus we call this quantity $\vec{u}$ the magnetic dipole moment. Unlike electric fields magnetic fields do not have 'charge 'counterparts. In other words there are no sources or sinks of magnetic fields, there can only be a dipole. Anything that can produce a magnetic field comes with both a source and a sink i.e. there is both a north pole and south pole. In many ways, the magnetic dipole is the fundamental unit that can produce a magnetic field.

Most elementary particles behave intrinsically as magnetic dipoles. For example, the electron itself behaves as a magnetic dipole and has a Spin Magnetic Dipole moment. This magnetic moment is intrinsic as the electron has neither an area A (it is a point object) nor does it spin around itself but is fundamental to the nature of the electron's existence.

We can generalize the magnetic moment for 'N' turns of the wire loop as,

$\mu = NiA$

The magnetic field lines of a current loop look similar to that of an idealized electric dipole:

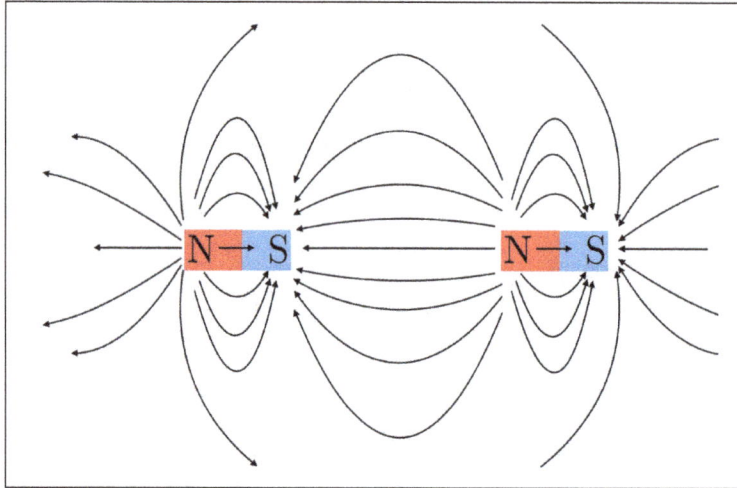

Magnetic field lines.

If you have ever broken a magnet into two parts, you would have found that each piece forms a new magnet. The new pieces also contain a north and a south pole. You just can't seem to be able to obtain just a North Pole.

Magnetic Dipole

Magnetic dipole is generally a tiny magnet of microscopic to subatomic dimensions, equivalent to a flow of electric charge around a loop. Electrons circulating around atomic nuclei, electrons spinning on their axes, and rotating positively charged atomic nuclei all are magnetic dipoles. The sum of these effects may cancel so that a given type of atom may not be a magnetic dipole. If they do not fully cancel, the atom is a permanent magnetic dipole, as are iron atoms. Many millions of iron atoms spontaneously locked into the same alignment to form a ferromagnetic domain also constitute a magnetic dipole. Magnetic compass needles and bar magnets are examples of macroscopic magnetic dipoles.

The strength of a magnetic dipole, called the magnetic dipole moment, may be thought of as a measure of a dipole's ability to turn itself into alignment with a given external magnetic field. In a uniform magnetic field, the magnitude of the dipole moment is proportional to the maximum amount of torque on the dipole, which occurs when the dipole is at right angles to the magnetic field. The magnetic dipole moment, often simply called the magnetic moment, may be defined then as the maximum amount of torque caused by magnetic force on a dipole that arises per unit value of surrounding magnetic field in vacuum.

When a magnetic dipole is considered as a current loop, the magnitude of the dipole moment is proportional to the current multiplied by the size of the enclosed area. The direction of the dipole moment, which may be represented mathematically as a vector, is perpendicularly away from the side of the surface enclosed by the counterclockwise path of positive charge flow. Considering the current loop as a tiny magnet, this vector corresponds to the direction from the south to the north pole. When free to rotate, dipoles align themselves so that their moments point predominantly in the direction of the external magnetic field. Nuclear and electron magnetic moments are quantized, which means that they may be oriented in space at only certain discrete angles with respect to the direction of the external field.

Magnetic dipole moments have dimensions of current times area or energy divided by magnetic flux density. In the metre–kilogram– second–ampere and SI systems, the specific unit for dipole moment is ampere-square metre. In the centimetre–gram–second electromagnetic system, the unit is the erg (unit of energy) per gauss (unit of magnetic flux density). One thousand ergs per gauss equal one ampere-square metre. A convenient unit for the magnetic dipole moment of electrons is the Bohr magneton (equivalent to 9.27×10^{-24} ampere–square metre). A similar unit for magnetic moments of nuclei, protons, and neutrons is the nuclear magneton (equivalent to 5.051×10^{-27} ampere–square metre).

References

- Lang, Kenneth R. (2006). A Companion to Astronomy and Astrophysics. Springer. P. 176. ISBN 9780387333670. Retrieved 19 April 2018

- Biot-savart-law: electrical4u.com, Retrieved 29 March, 2019

- Young, Hugh D.; Freedman, Roger A.; Ford, A. Lewis (2008). Sears and Zemansky's university physics : with modern physics. 2. Pearson Addison-Wesley. P. 918 - 919. ISBN 9780321501219

- Hu, Kaibo; Ma, Yicong; Xu, Jinchao (1 February 2017). "Stable finite element methods preserving $\nabla \cdot B = 0$ exactly for MHD models". Numerische Mathematik. 135 (2): 371–396. Doi:10.1007/s00211-016-0803-4. ISSN 0945-3245

- Magnetic-dipole-moment, physics: byjus.com, Retrieved 30 April, 2019

- Purcell, E. (2011). Electricity and Magnetism (2nd ed.). Cambridge University Press. Pp. 173–4. ISBN 978-110 7013605

- Magnetic-dipole, science: britannica.com, Retrieved 29 June, 2019

4
Electrodynamics

The study of the phenomena associated with charged bodies in motion, and fluctuating electric and magnetic fields is referred to as electrodynamics. There are various forces studied within this field such as Lorentz force, Electromotive force and Magnetomotive force. This chapter has been carefully written to provide an easy understanding of these aspects of electrodynamics.

Electrodynamics is the study of phenomena associated with charged bodies in motion and varying electric and magnetic fields; since a moving charge produces a magnetic field, electrodynamics is concerned with effects such as magnetism , electromagnetic radiation , and electromagnetic induction , including such practical applications as the electric generator and the electric motor. This area of electrodynamics, often known as classical electrodynamics, was first systematically explained by the physicist James Clerk Maxwell. Maxwell's equations, a set of differential equations, describe the phenomena of this area with great generality. A more recent development is quantum electrodynamics, which was formulated to explain the interaction of electromagnetic radiation with matter, to which the laws of the quantum theory apply. The physicists P. A. M. Dirac, W. Heisenberg, and W. Pauli were the pioneers in the formulation of quantum electrodynamics. When the velocities of the charged particles under consideration become comparable with the speed of light, corrections involving the theory of relativity must be made; this branch of the theory is called relativistic electrodynamics. It is applied to phenomena involved with particle accelerators and with electron tubes that are subject to high voltages and carry heavy currents.

Lorentz Force

In physics (specifically in electromagnetism) the Lorentz force (or electromagnetic force) is the combination of electric and magnetic force on a point charge due to electromagnetic fields. A particle of charge q moving with a velocity v in an electric field E and a magnetic field B experiences a force of,

$$F = qE + qv \times B$$

(in SI units). Variations on this basic formula describe the magnetic force on a current-carrying wire (sometimes called Laplace force), the electromotive force in a wire loop moving through a magnetic field (an aspect of Faraday's law of induction), and the force on a charged particle which might be traveling near the speed of light (relativistic form of the Lorentz force).

Historians suggest that the law is implicit in a paper by James Clerk Maxwell, published in 1865. Hendrik Lorentz arrived in a complete derivation in 1895, identifying the contribution of the electric force a few years after Oliver Heaviside correctly identified the contribution of the magnetic force.

Charged Particle

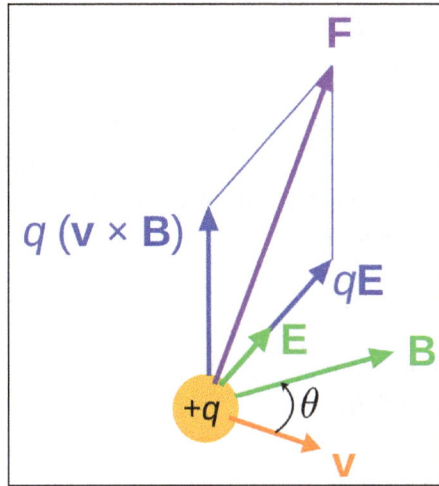

Lorentz force F on a charged particle (of charge q) in motion (instantaneous velocity v).
The E field and B field vary in space and time.

The force F acting on a particle of electric charge q with instantaneous velocity v, due to an external electric field E and magnetic field B, is given by (in SI units):

$$F = q(E + v \times B)$$

where × is the vector cross product (all boldface quantities are vectors). In terms of cartesian components, we have:

$$F_x = q(E_x + v_y B_z - v_z B_y),$$
$$F_y = q(E_y + v_z B_x - v_x B_z),$$
$$F_z = q(E_z + v_x B_y - v_y B_x).$$

In general, the electric and magnetic fields are functions of the position and time. Therefore, explicitly, the Lorentz force can be written as:

$$F(r,\dot{r},t,q) = q[E(r,t) + \dot{r} \times B(r,t)]$$

in which r is the position vector of the charged particle, t is time, and the overdot is a time derivative.

A positively charged particle will be accelerated in the *same* linear orientation as the E field, but will curve perpendicularly to both the instantaneous velocity vector v and the B field according to the right-hand rule (in detail, if the fingers of the right hand are extended to point in the direction of v and are then curled to point in the direction of B, then the extended thumb will point in the direction of F).

The term qE is called the electric force, while the term q(v × B) is called the magnetic force. According to some definitions, the term "Lorentz force" refers specifically to the formula for the magnetic force, with the *total* electromagnetic force (including the electric force) given some other (non-standard) name. This article will *not* follow this nomenclature: In what follows, the term "Lorentz force" will refer to the expression for the total force.

The magnetic force component of the Lorentz force manifests itself as the force that acts on a current-carrying wire in a magnetic field. In that context, it is also called the Laplace force.

The Lorentz force is a force exerted by the electromagnetic field on the charged particle, that is, it is the rate at which linear momentum is transferred from the electromagnetic field to the particle. Associated with it is the power which is the rate at which energy is transferred from the electromagnetic field to the particle. That power is,

$$v \cdot F = q\, v \cdot E.$$

Notice that the magnetic field does not contribute to the power because the magnetic force is always perpendicular to the velocity of the particle.

Continuous Charge Distribution

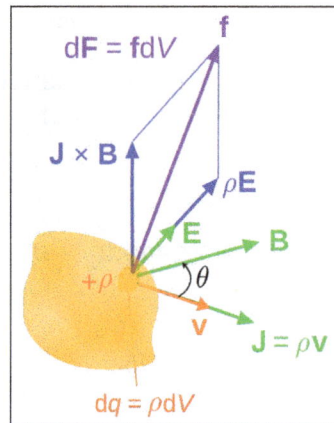

Lorentz force (per unit 3-volume) f on a continuous charge distribution (charge density ρ) in motion. The 3-current density J corresponds to the motion of the charge element dq in volume element dV and varies throughout the continuum.

For a continuous charge distribution in motion, the Lorentz force equation becomes:

$$dF = dq \left(E + v \times B \right)$$

where dF is the force on a small piece of the charge distribution with charge dq. If both sides of this equation are divided by the volume of this small piece of the charge distribution dV, the result is:

$$f = \rho \left(E + v \times B \right)$$

where f is the force density (force per unit volume) and ρ is the charge density (charge per unit volume). Next, the current density corresponding to the motion of the charge continuum is,

$$J = \rho v$$

so the continuous analogue to the equation is,

$$f = \rho E + J \times B$$

The total force is the volume integral over the charge distribution:

$$F = \iiint (\rho E + J \times B)\, dV.$$

By eliminating ρ and J, using Maxwell's equations, and manipulating using the theorems of vector calculus, this form of the equation can be used to derive the Maxwell stress tensor σ, in turn this can be combined with the Poynting vector S to obtain the electromagnetic stress–energy tensor T used in general relativity.

In terms of σ and S, another way to write the Lorentz force (per unit volume) is,

$$f = \nabla \cdot \sigma - \frac{1}{c^2} \frac{\partial S}{\partial t}$$

where c is the speed of light and $\nabla \cdot$ denotes the divergence of a tensor field. Rather than the amount of charge and its velocity in electric and magnetic fields, this equation relates the energy flux (flow of energy per unit time per unit distance) in the fields to the force exerted on a charge distribution.

The density of power associated with the Lorentz force in a material medium is,

$$J \cdot E.$$

If we separate the total charge and total current into their free and bound parts, we get that the density of the Lorentz force is,

$$f = (\rho_f - \nabla \cdot P)E + (J_f + \nabla \times M + \frac{\partial P}{\partial t}) \times B.$$

where: ρ_f is the density of free charge; P is the polarization density; J_f is the density of free current; and M is the magnetization density. In this way, the Lorentz force can explain the torque applied to a permanent magnet by the magnetic field. The density of the associated power is

$$\left(J_f + \nabla \times M + \frac{\partial P}{\partial t} \right) \cdot E.$$

Equation in CGS Units

The above-mentioned formulae use SI units which are the most common among experimentalists, technicians, and engineers. In cgs-Gaussian units, which are somewhat more common among the-oretical physicists as well as condensed matter experimentalists, one has instead,

$$F = q_{cgs}\left(E_{cgs} + \frac{v}{c} \times B_{cgs}\right).$$

where c is the speed of light. Although this equation looks slightly different, it is completely equiv-alent, since one has the following relations:

$$q_{cgs} = \frac{q_{SI}}{\sqrt{4\pi\epsilon_0}}, \quad E_{cgs} = \sqrt{4\pi\epsilon_0}\, E_{SI}, \quad B_{cgs} = \sqrt{4\pi/\mu_0}\, B_{SI}, \quad c = \frac{1}{\sqrt{\epsilon_0\mu_0}}.$$

where ϵ_0 is the vacuum permittivity and μ_0 the vacuum permeability. In practice, the subscripts "cgs" and "SI" are always omitted, and the unit system has to be assessed from context.

Trajectories of Particles Due to the Lorentz Force

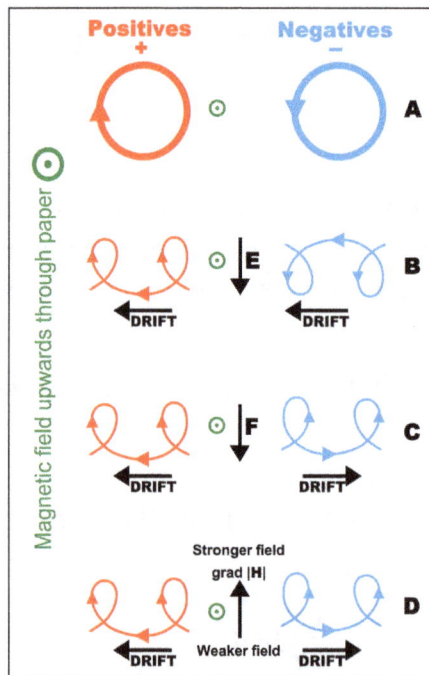

Charged Particle Drifts in a homogeneous magnetic field. (A) No disturbing force (B) With an electric field, E (C) With an independent force, F (e.g. gravity) (D) In an inhomogeneous magnetic field, grad H.

In many cases of practical interest, the motion in a magnetic field of an electrically charged parti-cle (such as an electron or ion in a plasma) can be treated as the superposition of a relatively fast circular motion around a point called the guiding center and a relatively slow drift of this point. The drift speeds may differ for various species depending on their charge states, masses, or tem-peratures, possibly resulting in electric currents or chemical separation.

Significance of the Lorentz Force

While the modern Maxwell's equations describe how electrically charged particles and currents or moving charged particles give rise to electric and magnetic fields, the Lorentz force law completes that picture by describing the force acting on a moving point charge q in the presence of electromagnetic fields. The Lorentz force law describes the effect of E and B upon a point charge, but such electromagnetic forces are not the entire picture. Charged particles are possibly coupled to other forces, notably gravity and nuclear forces. Thus, Maxwell's equations do not stand separate from other physical laws, but are coupled to them via the charge and current densities. The response of a point charge to the Lorentz law is one aspect; the generation of E and B by currents and charges is another.

In real materials the Lorentz force is inadequate to describe the collective behavior of charged particles, both in principle and as a matter of computation. The charged particles in a material medium not only respond to the E and B fields but also generate these fields. Complex transport equations must be solved to determine the time and spatial response of charges, for example, the Boltzmann equation or the Fokker–Planck equation or the Navier–Stokes equations.

Lorentz Force Law as the Definition of E and B

In many textbook treatments of classical electromagnetism, the Lorentz force Law is used as the definition of the electric and magnetic fields E and B. To be specific, the Lorentz force is understood to be the following empirical statement:

> The electromagnetic force F on a test charge at a given point and time is a certain function of its charge q and velocity v, which can be parameterized by exactly two vectors E and B, in the functional form:

$$F = q(E + v \times B)$$

This is valid, even for particles approaching the speed of light (that is, magnitude of v = |v| ≈ c). So the two vector fields E and B are thereby defined throughout space and time, and these are called the "electric field" and "magnetic field". The fields are defined everywhere in space and time with respect to what force a test charge would receive regardless of whether a charge is present to experience the force.

As a definition of E and B, the Lorentz force is only a definition in principle because a real particle (as opposed to the hypothetical "test charge" of infinitesimally-small mass and charge) would generate its own finite E and B fields, which would alter the electromagnetic force that it experiences. In addition, if the charge experiences acceleration, as if forced into a curved trajectory by some external agency, it emits radiation that causes braking of its motion. These effects occur through both a direct effect (called the radiation reaction force) and indirectly (by affecting the motion of nearby charges and currents). Moreover, net force must include gravity, electroweak, and any other forces aside from electromagnetic force.

Force on a Current-carrying Wire

When a wire carrying an electric current is placed in a magnetic field, each of the moving charges, which comprise the current, experiences the Lorentz force, and together they can create a macroscopic force on the wire (sometimes called the Laplace force). By combining the Lorentz force law above with the definition of electric current, the following equation results, in the case of a straight, stationary wire:

$$F = I\ell \times B$$

where ℓ is a vector whose magnitude is the length of wire, and whose direction is along the wire, aligned with the direction of conventional current charge flow I.

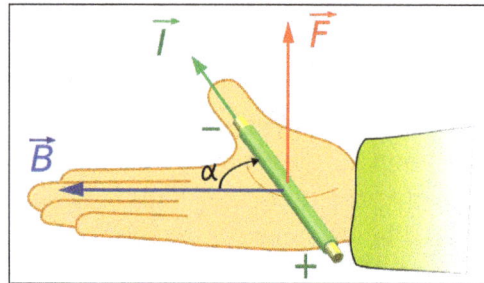

Right-hand rule for a current-carrying wire in a magnetic field B.

If the wire is not straight but curved, the force on it can be computed by applying this formula to each infinitesimal segment of wire dℓ, then adding up all these forces by integration. Formally, the net force on a stationary, rigid wire carrying a steady current I is

$$F = I\int d\ell \times B$$

This is the net force. In addition, there will usually be torque, plus other effects if the wire is not perfectly rigid.

One application of this is Ampère's force law, which describes how two current-carrying wires can attract or repel each other, since each experiences a Lorentz force from the other's magnetic field.

EMF

The magnetic force ($q\mathbf{v} \times B$) component of the Lorentz force is responsible for *motional* electromotive force (or motional EMF), the phenomenon underlying many electrical generators. When a conductor is moved through a magnetic field, the magnetic field exerts opposite forces on electrons and nuclei in the wire, and this creates the EMF. The term "motional EMF" is applied to this phenomenon, since the EMF is due to the motion of the wire.

In other electrical generators, the magnets move, while the conductors do not. In this case, the EMF is due to the electric force (qE) term in the Lorentz Force equation. The electric field in question is created by the changing magnetic field, resulting in an induced EMF, as described by the Maxwell–Faraday equation (one of the four modern Maxwell's equations).

Both of these EMFs, despite their apparently distinct origins, are described by the same equation, namely, the EMF is the rate of change of magnetic flux through the wire. (This is Faraday's law of induction) Einstein's special theory of relativity was partially motivated by the desire to better understand this link between the two effects. In fact, the electric and magnetic fields are different facets of the same electromagnetic field, and in moving from one inertial frame to another, the solenoidal vector field portion of the E-field can change in whole or in part to a B-field *vice versa*.

Lorentz Force and Faraday's Law of Induction

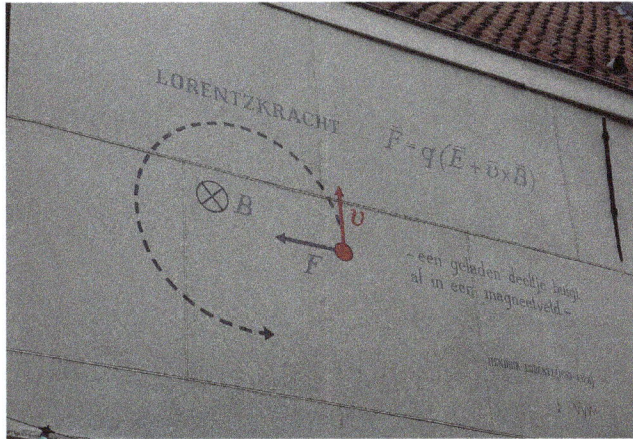

Lorentz force -image on a wall in Leiden.

Given a loop of wire in a magnetic field, Faraday's law of induction states the induced electromotive force (EMF) in the wire is:

$$\mathcal{E} = -\frac{d\Phi_B}{dt}$$

Where,

$$\Phi_B = \iint_{\Sigma(t)} dA \cdot B(r,t)$$

is the magnetic flux through the loop, B is the magnetic field, $\Sigma(t)$ is a surface bounded by the closed contour $\partial\Sigma(t)$, at all at time t, dA is an infinitesimal vector area element of $\Sigma(t)$ (magnitude is the area of an infinitesimal patch of surface, direction is orthogonal to that surface patch).

The sign of the EMF is determined by Lenz's law. Note that this is valid for not only a stationary wire – but also for a moving wire.

From Faraday's law of induction (that is valid for a moving wire, for instance in a motor) and the Maxwell Equations, the Lorentz Force can be deduced. The reverse is also true, the Lorentz force and the Maxwell Equations can be used to derive the Faraday Law.

Let $\Sigma(t)$ be the moving wire, moving together without rotation and with constant velocity v and $\Sigma(t)$ be the internal surface of the wire. The EMF around the closed path $\partial\Sigma(t)$ is given by:

$$E = \oint_{\partial\Sigma(t)} d\ell \cdot F / q$$

Where,

$$E = F / q$$

is the electric field and $d\ell$ is an infinitesimal vector element of the contour $\partial\Sigma(t)$.

NB: Both dℓ and dA have a sign ambiguity; to get the correct sign, the right-hand rule is used.

The above result can be compared with the version of Faraday's law of induction that appears in the modern Maxwell's equations, called here the Maxwell–Faraday equation:

$$\nabla \times E = -\frac{\partial B}{\partial t}.$$

The Maxwell–Faraday equation also can be written in an integral form using the Kelvin–Stokes theorem.

So we have, the Maxwell Faraday equation:

$$\oint_{\partial \Sigma(t)} d\ell \cdot E(r,t) = -\iint_{\Sigma(t)} dA \cdot \frac{d\,B(r,t)}{dt}$$

and the Faraday Law,

$$\oint_{\partial \Sigma(t)} d\ell \cdot F / q(r,t) = -\frac{d}{dt} \iint_{\Sigma(t)} dA \cdot B(r,t)$$

The two are equivalent if the wire is not moving. Using the Leibniz integral rule and that div B = 0, results in,

$$\oint_{\partial \Sigma(t)} d\ell \cdot F / q(r,t) = -\iint_{\Sigma(t)} dA \cdot \frac{\partial}{\partial t} B(r,t) + \oint_{\partial \Sigma(t)} v \times B\, d\ell$$

and using the Maxwell Faraday equation,

$$\oint_{\partial \Sigma(t)} d\ell \cdot F / q(r,t) = \oint_{\partial \Sigma(t)} d\ell \cdot E(r,t) + \oint_{\partial \Sigma(t)} v \times B(r,t)\, d\ell$$

since this is valid for any wire position it implies that,

$$F = q\, E(r,t) + q\, v \times B(r,t)$$

Faraday's law of induction holds whether the loop of wire is rigid and stationary, or in motion or in process of deformation, and it holds whether the magnetic field is constant in time or changing. However, there are cases where Faraday's law is either inadequate or difficult to use, and application of the underlying Lorentz force law is necessary.

If the magnetic field is fixed in time and the conducting loop moves through the field, the magnetic flux Φ_B linking the loop can change in several ways. For example, if the B-field varies with position, and the loop moves to a location with different B-field, Φ_B will change. Alternatively, if the loop changes orientation with respect to the B-field, the B · dA differential element will change because of the different angle between B and dA, also changing Φ_B. As a third example, if a portion of the circuit is swept through a uniform, time-independent B-field, and another portion of the circuit is held stationary, the flux linking the entire closed circuit can change due to the shift in relative position of the circuit's component parts with time (surface $\partial \Sigma(t)$

time-dependent). In all three cases, Faraday's law of induction then predicts the EMF generated by the change in Φ_B.

Note that the Maxwell Faraday's equation implies that the Electric Field E is non conservative when the Magnetic Field B varies in time, and is not expressible as the gradient of a scalar field, and not subject to the gradient theorem since its rotational is not zero.

Lorentz Force in Terms of Potentials

The E and B fields can be replaced by the magnetic vector potential A and (scalar) electrostatic potential ϕ by,

$$E = -\nabla\phi - \frac{\partial A}{\partial t}$$

$$B = \nabla \times A$$

where ∇ is the gradient, $\nabla \cdot$ is the divergence, $\nabla \times$ is the curl.

The force becomes,

$$F = q\left[-\nabla\phi - \frac{\partial A}{\partial t} + v \times (\nabla \times A)\right]$$

and using an identity for the triple product simplifies to,

$$F = q\left[-\nabla\phi - \frac{\partial A}{\partial t} + \nabla(v \cdot A) - (v \cdot \nabla)A\right]$$

(v has no dependence on position, so there's no need to use Feynman's subscript notation). Using the chain rule, the total derivative of A is:

$$\frac{dA}{dt} = \frac{\partial A}{\partial t} + (v \cdot \nabla)A$$

so the above expression can be rewritten as:

$$F = q\left[-\nabla(\phi - v \cdot A) - \frac{dA}{dt}\right].$$

With $v = \dot{x}$, we can put the equation into the convenient Euler–Lagrange form:

$$F = q\left[-\nabla_x(\phi - \dot{x} \cdot A) + \frac{d}{dt}\nabla_{\dot{x}}(\phi - \dot{x} \cdot A)\right]$$

Where,

$$\nabla_x = \hat{x}\frac{\partial}{\partial x} + \hat{y}\frac{\partial}{\partial y} + \hat{z}\frac{\partial}{\partial z}$$

And

$$\nabla_{\dot{x}} = \hat{x}\frac{\partial}{\partial\dot{x}} + \hat{y}\frac{\partial}{\partial\dot{y}} + \hat{z}\frac{\partial}{\partial\dot{z}}.$$

Lorentz Force and Analytical Mechanics

The Lagrangian for a charged particle of mass m and charge q in an electromagnetic field equivalently describes the dynamics of the particle in terms of its *energy*, rather than the force exerted on it. The classical expression is given by:

$$L = \frac{m}{2}\dot{r}\cdot\dot{r} + qA\cdot\dot{r} - q\phi$$

where A and ϕ are the potential fields as above. Using Lagrange's equations, the equation for the Lorentz force can be obtained.

Derivation of Lorentz force from classical Lagrangian (SI units).

For an A field, a particle moving with velocity v = $\dot{r}$ has potential momentum $qA(r,t)$, so its potential energy is $qA(r,t)\cdot\dot{r}$. For a ϕ field, the particle's potential energy is $q\phi(r,t)$.

The total potential energy is then:

$$V = q\phi - qA\cdot\dot{r}$$

and the kinetic energy is:

$$T = \frac{m}{2}\dot{r}\cdot\dot{r}$$

hence, the Lagrangian:

$$L = T - V = \frac{m}{2}\dot{r}\cdot\dot{r} + qA\cdot\dot{r} - q\phi$$

$$L = \frac{m}{2}(\dot{x}^2 + \dot{y}^2 + \dot{z}^2) + q(\dot{x}A_x + \dot{y}A_y + \dot{z}A_z) - q\phi$$

Lagrange's equations are,

$$\frac{d}{dt}\frac{\partial L}{\partial\dot{x}} = \frac{\partial L}{\partial x}$$

(same for y and z). So, calculating the partial derivatives:

$$\frac{d}{dt}\frac{\partial L}{\partial\dot{x}} = m\ddot{x} + q\frac{dA_x}{dt}$$

$$= m\ddot{x} + \frac{q}{dt}\left(\frac{\partial A_x}{\partial t}dt + \frac{\partial A_x}{\partial x}dx + \frac{\partial A_x}{\partial y}dy + \frac{\partial A_x}{\partial z}dz\right)$$

$$= m\ddot{x} + q\left(\frac{\partial A_x}{\partial t} + \frac{\partial A_x}{\partial x}\dot{x} + \frac{\partial A_x}{\partial y}\dot{y} + \frac{\partial A_x}{\partial z}\dot{z}\right)$$

$$\frac{\partial L}{\partial x} = -q\frac{\partial \phi}{\partial x} + q\left(\frac{\partial A_x}{\partial x}\dot{x} + \frac{\partial A_y}{\partial x}\dot{y} + \frac{\partial A_z}{\partial x}\dot{z}\right)$$

equating and simplifying:

$$m\ddot{x} + q\left(\frac{\partial A_x}{\partial t} + \frac{\partial A_x}{\partial x}\dot{x} + \frac{\partial A_x}{\partial y}\dot{y} + \frac{\partial A_x}{\partial z}\dot{z}\right) = -q\frac{\partial \phi}{\partial x} + q\left(\frac{\partial A_x}{\partial x}\dot{x} + \frac{\partial A_y}{\partial x}\dot{y} + \frac{\partial A_z}{\partial x}\dot{z}\right)$$

$$F_x = -q\left(\frac{\partial \phi}{\partial x} + \frac{\partial A_x}{\partial t}\right) + q\left[\dot{y}\left(\frac{\partial A_y}{\partial x} - \frac{\partial A_x}{\partial y}\right) + \dot{z}\left(\frac{\partial A_z}{\partial x} - \frac{\partial A_x}{\partial z}\right)\right]$$

$$= qE_x + q[\dot{y}(\nabla \times A)_z - \dot{z}(\nabla \times A)_y]$$

$$= qE_x + q[\dot{r} \times (\nabla \times A)]_x$$

$$= qE_x + q(\dot{r} \times B)_x$$

and similarly for the y and z directions. Hence, the force equation is:

$$F = q(E + \dot{r} \times B)$$

The potential energy depends on the velocity of the particle, so the force is velocity dependent, so it is not conservative. The relativistic Lagrangian is:

$$L = -mc^2\sqrt{1 - \left(\frac{\dot{r}}{c}\right)^2} + qA(r) \cdot \dot{r} - q\phi(r)$$

The action is the relativistic arclength of the path of the particle in space time, minus the potential energy contribution, plus an extra contribution which quantum mechanically is an extra phase a charged particle gets when it is moving along a vector potential.

Derivation of Lorentz Force from Relativistic Lagrangian (SI Units)

The equations of motion derived by extremizing the action:

$$\frac{dP}{dt} = \frac{\partial L}{\partial r} = q\frac{\partial A}{\partial r} \cdot \dot{r} - q\frac{\partial \phi}{\partial r}$$

$$P - qA - \frac{m\dot{r}}{\sqrt{1 - \left(\frac{\dot{r}}{c}\right)^2}}$$

are the same as Hamilton's equations of motion:

$$\frac{dr}{dt} = \frac{\partial}{\partial p}\left(\sqrt{(P - qA)^2 + (mc^2)^2} + q\phi\right)$$

$$\frac{dp}{dt} = -\frac{\partial}{\partial r}\left(\sqrt{(P - qA)^2 + (mc^2)^2} + q\phi\right)$$

both are equivalent to the noncanonical form:

$$\frac{d}{dt}\left(\frac{m\dot{r}}{\sqrt{1-\left(\dfrac{\dot{r}}{c}\right)^2}} \right) = q\left(E + \dot{r} \times B\right).$$

This formula is the Lorentz force, representing the rate at which the EM field adds relativistic momentum to the particle.

Relativistic Form of the Lorentz Force

Covariant Form of the Lorentz Force

Field Tensor:

Using the metric signature (1, −1, −1, −1), the Lorentz force for a charge q can be written in covariant form:

$$\frac{dp^\alpha}{d\tau} = qF^{\alpha\beta}U_\beta$$

where p^α is the four-momentum, defined as,

$$p^\alpha = \left(p_0, p_1, p_2, p_3\right) = \left(\gamma mc, p_x, p_y, p_z\right),$$

τ the proper time of the particle, $F^{\alpha\beta}$ the contravariant electromagnetic tensor,

$$F^{\alpha\beta} = \begin{pmatrix} 0 & -E_x/c & -E_y/c & -E_z/c \\ E_x/c & 0 & -B_z & B_y \\ E_y/c & B_z & 0 & -B_x \\ E_z/c & -B_y & B_x & 0 \end{pmatrix}$$

and U is the covariant 4-velocity of the particle, defined as:

$$U_\beta = \left(U_0, U_1, U_2, U_3\right) = \gamma\left(c, -v_x, -v_y, -v_z\right),$$

in which,

$$\gamma(v) = \frac{1}{\sqrt{1-\dfrac{v^2}{c^2}}} = \frac{1}{\sqrt{1-\dfrac{v_x^2+v_y^2+v_z^2}{c^2}}}$$

is the Lorentz factor.

The fields are transformed to a frame moving with constant relative velocity by:

$$F'^{\mu\nu} = \Lambda^{\mu}{}_{\alpha}\Lambda^{\nu}{}_{\beta}F^{\alpha\beta},$$

where $\Lambda^{\mu}{}_{\alpha}$ is the Lorentz transformation tensor.

Translation to Vector Notation

The $\alpha = 1$ component (x-component) of the force is,

$$\frac{dp^1}{d\tau} = qU_{\beta}F^{1\beta} = q\left(U_0 F^{10} + U_1 F^{11} + U_2 F^{12} + U_3 F^{13}\right).$$

Substituting the components of the covariant electromagnetic tensor F yields,

$$\frac{dp^1}{d\tau} = q\left[U_0\left(\frac{E_x}{c}\right) + U_2(-B_z) + U_3(B_y)\right].$$

Using the components of covariant four-velocity yields,

$$\frac{dp^1}{d\tau} = q\gamma\left[c\left(\frac{E_x}{c}\right) + (-v_y)(-B_z) + (-v_z)(B_y)\right]$$
$$= q\gamma\left(E_x + v_y B_z - v_z B_y\right)$$
$$= q\gamma\left[E_x + (v \times B)_x\right].$$

The calculation for $\alpha = 2, 3$ (force components in the y and z directions) yields similar results, so collecting the 3 equations into one:

$$\frac{dp}{d\tau} = q\gamma\left(E + v \times B\right),$$

and since differentials in coordinate time dt and proper time $d\tau$ are related by the Lorentz factor,

$$dt = \gamma(v)d\tau,$$

so we arrive at,

$$\frac{dp}{dt} = q\left(E + v \times B\right).$$

This is precisely the Lorentz force law, however, it is important to note that p is the relativistic expression,

$$p = \gamma(v)m_0 v.$$

Lorentz Force in General Relativity

In the general theory of relativity the equation of motion for a particle with mass m and charge e, moving in a space with metric tensor g_{ab} and electromagnetic field F_{ab}, is given as,

$$m\frac{du_c}{ds} - m\frac{1}{2}g_{ab,c}u^a u^b = eF_{cb}u^b,$$

Where $u^a = dx^a / ds$ (dx^a is taken along the trajectory), $g_{ab,c} = \partial g_{ab}/\partial x^c$, and $ds^2 = g_{ab}dx^a dx^b$.

The equation can also be written as,

$$m\frac{du_c}{ds} - m\Gamma_{abc}u^a u^b = eF_{cb}u^b,$$

Where Γ_{abc} is the Christoffel symbol (of the torsion-free metric connection in general relativity), or as,

$$m\frac{Du_c}{ds} = eF_{cb}u^b,$$

where D is the covariant differential in general relativity (metric, torsion-free).

Electromotive Force

Electromotive force, abbreviated emf (denoted ε and measured in volts), is the electrical action produced by a non-electrical source. A device that converts other forms of energy into electrical energy (a "transducer"), such as a battery (converting chemical energy) or generator (converting mechanical energy), provides an emf as its output. Sometimes an analogy to water "pressure" is used to describe electromotive force. (The word "force" in this case is not used to mean force of interaction between bodies, as may be measured in pounds or newtons.)

In electromagnetic induction, emf can be defined around a closed loop of conductor as the electromagnetic work that would be done on an electric charge (an electron in this instance) if it travels once around the loop. For a time-varying magnetic flux linking a loop, the electric potential scalar field is not defined due to a circulating electric vector field, but an emf nevertheless does work that can be measured as a virtual electric potential around the loop.

In the case of a two-terminal device (such as an electrochemical cell) which is modeled as a Thévenin's equivalent circuit, the equivalent emf can be measured as the open-circuit potential difference or "voltage" between the two terminals. This potential difference can drive an electric current if an external circuit is attached to the terminals.

Devices that can provide emf include electrochemical cells, thermoelectric devices, solar cells, photodiodes, electrical generators, transformers and even Van de Graaff generators. In nature, emf is generated whenever magnetic field fluctuations occur through a surface. The shifting of the

Earth's magnetic field during a geomagnetic storm induces currents in the electrical grid as the lines of the magnetic field are shifted about and cut across the conductors.

In the case of a battery, the charge separation that gives rise to a voltage difference between the terminals is accomplished by chemical reactions at the electrodes that convert chemical potential energy into electromagnetic potential energy. A voltaic cell can be thought of as having a "charge pump" of atomic dimensions at each electrode, that is:

A source of emf can be thought of as a kind of *charge pump* that acts to move positive charge from a point of low potential through its interior to a point of high potential. ... By chemical, mechanical or other means, the source of emf performs work dW on that charge to move it to the high potential terminal. The emf E of the source is defined as the work dW done per charge dq: $E = dW/dq$.

In the case of an electrical generator, a time-varying magnetic field inside the generator creates an electric field via electromagnetic induction, which in turn creates a voltage difference between the generator terminals. Charge separation takes place within the generator, with electrons flowing away from one terminal and toward the other, until, in the open-circuit case, sufficient electric field builds up to make further charge separation impossible. Again, the emf is countered by the electrical voltage due to charge separation. If a load is attached, this voltage can drive a current. The general principle governing the emf in such electrical machines is Faraday's law of induction.

Notation and Units of Measurement

Electromotive force is often denoted by ε or E (script capital E, Unicode U+2130).

In a device without internal resistance, if an electric charge Q passes through that device, and gains an energy W, the net emf for that device is the energy gained per unit charge, or W/Q. Like other measures of energy per charge, emf uses the SI unit volt, which is equivalent to a joule per coulomb.

Electromotive force in electrostatic units is the statvolt (in the centimeter gram second system of units equal in amount to an erg per electrostatic unit of charge).

Formal Definitions

Inside a source of emf that is open-circuited, the conservative electrostatic field created by separation of charge exactly cancels the forces producing the emf. Thus, the emf has the same value but opposite sign as the integral of the electric field aligned with an internal path between two terminals A and B of a source of emf in open-circuit condition (the path is taken from the negative terminal to the positive terminal to yield a positive emf, indicating work done on the electrons moving in the circuit). Mathematically:

$$E = -\int_{A}^{B} E_{cs} \cdot d\ell \,,$$

where E_{cs} is the conservative electrostatic field created by the charge separation associated with the emf, $d\ell$ is an element of the path from terminal A to terminal B, and '·' denotes the vector dot product. This equation applies only to locations A and B that are terminals, and does not apply to paths between points A and B with portions outside the source of emf. This equation involves the electrostatic electric field due to charge separation E_{cs} and does not involve (for example) any non-conservative component of electric field due to Faraday's law of induction.

In the case of a closed path in the presence of a varying magnetic field, the integral of the electric field around a closed loop may be nonzero; one common application of the concept of emf, known as "*induced emf*" is the voltage induced in such a loop. The "*induced emf*" around a stationary closed path C is:

$$E = \oint_C E \cdot d\ell \,,$$

where E is the entire electric field, conservative and non-conservative, and the integral is around an arbitrary but stationary closed curve C through which there is a varying magnetic field. The electrostatic field does not contribute to the net emf around a circuit because the electrostatic portion of the electric field is conservative (i.e., the work done against the field around a closed path is zero, which is valid, as long as the circuit elements remain at rest and radiation is ignored).

This definition can be extended to arbitrary sources of emf and moving paths C:

$$E = \oint_C [E + v \times B] \cdot d\ell$$
$$+ \frac{1}{q} \oint_C \text{Effective chemical forces} \cdot d\ell$$
$$+ \frac{1}{q} \oint_C \text{Effective thermal forces} \cdot d\ell,$$

which is a conceptual equation mainly, because the determination of the "effective forces" is difficult.

Voltage Difference

An electrical voltage difference is sometimes called an emf. The points below illustrate the more formal usage, in terms of the distinction between emf and the voltage it generates:

1. For a circuit as a whole, such as one containing a resistor in series with a voltaic cell, electrical voltage does not contribute to the overall emf, because the voltage difference on going around a circuit is zero. (The ohmic IR voltage drop plus the applied electrical voltage sum to zero. See Kirchhoff's voltage law). The emf is due solely to the chemistry in the battery that causes charge separation, which in turn creates an electrical voltage that drives the current.

2. For a circuit consisting of an electrical generator that drives current through a resistor, the emf is due solely to a time-varying magnetic field within the generator that generates an electrical voltage that in turn drives the current. (The ohmic IR drop plus the applied electrical voltage again is zero.)

3. A transformer coupling two circuits may be considered a source of emf for one of the circuits, just as if it were caused by an electrical generator; this example illustrates the origin of the term "transformer emf".

4. A photodiode or solar cell may be considered as a source of emf, similar to a battery, resulting in an electrical voltage generated by charge separation driven by light rather than chemical reaction.

5. Other devices that produce emf are fuel cells, thermocouples, and thermopiles.

In the case of an open circuit, the electric charge that has been separated by the mechanism generating the emf creates an electric field opposing the separation mechanism. For example, the chemical reaction in a voltaic cell stops when the opposing electric field at each electrode is strong enough to arrest the reactions. A larger opposing field can reverse the reactions in what are called *reversible* cells.

The electric charge that has been separated creates an electric potential difference that can be measured with a voltmeter between the terminals of the device. The magnitude of the emf for the battery (or other source) is the value of this 'open circuit' voltage. When the battery is charging or discharging, the emf itself cannot be measured directly using the external voltage because some voltage is lost inside the source. It can, however, be inferred from a measurement of the current I and voltage difference V, provided that the internal resistance r already has been measured: $\varepsilon = V + Ir$.

Magnetomotive Force

In physics, the magnetomotive force (mmf) is a quantity appearing in the equation for the magnetic flux in a magnetic circuit, often called Ohm's law for magnetic circuits. It is the property of certain substances or phenomena that give rise to magnetic fields:

$$\mathcal{F} = \Phi\mathcal{R},$$

where Φ is the magnetic flux and R is the reluctance of the circuit. It can be seen that the magnetomotive force plays a role in this equation analogous to the voltage V in Ohm's law: $V = IR$, since it is the cause of magnetic flux in a magnetic circuit:

1. $F = NI$

   where N is the number of turns in the coil and I is the electric current through the circuit.

2. $F = \Phi R$

   where Φ is the magnetic flux and R is the magnetic reluctance

3. $F = HL$

   where H is the magnetizing force (the strength of the magnetizing field) and L is the mean length of a solenoid or the circumference of a toroid.

Units

The SI unit of mmf is the ampere, the same as the unit of current (analogously the units of emf and voltage are both the volt). Informally, and frequently, this unit is stated as the ampere-turn to avoid confusion with current. This was the unit name in the MKS system. Occasionally, the cgs system unit of the gilbert may also be encountered.

Electromagnetic Induction

Electromagnetic or magnetic induction is the production of an electromotive force (i.e., voltage) across an electrical conductor in a changing magnetic field.

Michael Faraday is generally credited with the discovery of induction in 1831, and James Clerk Maxwell mathematically described it as Faraday's law of induction. Lenz's law describes the direction of the induced field. Faraday's law was later generalized to become the Maxwell–Faraday equation, one of the four Maxwell equations in his theory of electromagnetism.

Electromagnetic induction has found many applications, including electrical components such as inductors and transformers, and devices such as electric motors and generators.

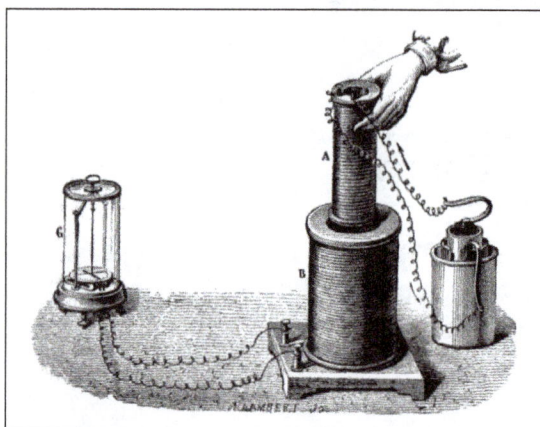

Faraday's experiment showing induction between coils of wire: The liquid battery *(right)* provides a current that flows through the small coil *(A)*, creating a magnetic field. When the coils are stationary, no current is induced. But when the small coil is moved in or out of the large coil *(B)*, the magnetic flux through the large coil changes, inducing a current which is detected by the galvanometer *(G)*.

Theory

Faraday's Law of Induction and Lenz's Law

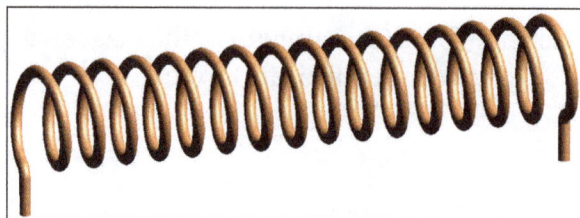

A solenoid

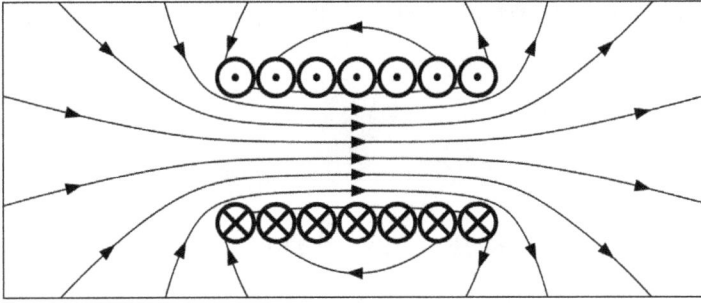

The longitudinal cross section of a solenoid with a constant electrical current running through it. The magnetic field lines are indicated, with their direction shown by arrows. The magnetic flux corresponds to the 'density of field lines'. The magnetic flux is thus densest in the middle of the solenoid, and weakest outside of it.

Faraday's law of induction makes use of the magnetic flux Φ_B through a region of space enclosed by a wire loop. The magnetic flux is defined by a surface integral:

$$\Phi_B = \int_\Sigma B \cdot dA ,$$

where dA is an element of the surface Σ enclosed by the wire loop, B is the magnetic field. The dot product B·dA corresponds to an infinitesimal amount of magnetic flux. In more visual terms, the magnetic flux through the wire loop is proportional to the number of magnetic flux lines that pass through the loop.

When the flux through the surface changes, Faraday's law of induction says that the wire loop acquires an electromotive force (EMF). The most widespread version of this law states that the induced electromotive force in any closed circuit is equal to the rate of change of the magnetic flux enclosed by the circuit:

$$\mathcal{E} = -\frac{d\Phi_B}{dt} ,$$

where $\mathcal{E}$ is the EMF and Φ_B is the magnetic flux. The direction of the electromotive force is given by Lenz's law which states that an induced current will flow in the direction that will oppose the change which produced it. This is due to the negative sign in the previous equation. To increase the generated EMF, a common approach is to exploit flux linkage by creating a tightly wound coil of wire, composed of N identical turns, each with the same magnetic flux going through them. The resulting EMF is then N times that of one single wire.

$$\mathcal{E} = -N\frac{d\Phi_B}{dt}$$

Generating an EMF through a variation of the magnetic flux through the surface of a wire loop can be achieved in several ways:

- The magnetic field B changes (e.g. an alternating magnetic field, or moving a wire loop towards a bar magnet where the B field is stronger),

- The wire loop is deformed and the surface Σ changes,

- The orientation of the surface dA changes (e.g. spinning a wire loop into a fixed magnetic field),

- Any combination of the above.

Maxwell–Faraday Equation

In general, the relation between the EMF $\mathcal{E}$ in a wire loop encircling a surface Σ, and the electric field E in the wire is given by,

$$E = \oint_{\partial\Sigma} E \cdot d\ell$$

where $d\ell$ is an element of contour of the surface Σ, combining this with the definition of flux,

$$\Phi_B = \int_\Sigma B \cdot dA \,,$$

we can write the integral form of the Maxwell–Faraday equation,

$$\oint_{\partial\Sigma} E \cdot d\ell = -\frac{d}{dt}\int_\Sigma B \cdot dA$$

It is one of the four Maxwell's equations, and therefore plays a fundamental role in the theory of classical electromagnetism.

Faraday's Law and Relativity

Faraday's law describes two different phenomena: the motional EMF generated by a magnetic force on a moving wire, and the transformer EMF this is generated by an electric force due to a changing magnetic field (due to the differential form of the Maxwell–Faraday equation). James Clerk Maxwell drew attention to the separate physical phenomena in 1861. This is believed to be a unique example in physics of where such a fundamental law is invoked to explain two such different phenomena.

Einstein noticed that the two situations both corresponded to a relative movement between a conductor and a magnet, and the outcome was unaffected by which one was moving. This was one of the principal paths that led him to develop special relativity.

Applications

The principles of electromagnetic induction are applied in many devices and systems, including:

- Current clamp

- Electric generators

- Electromagnetic forming

- Graphics tablet

- Hall effect meters

- Induction cooking

- Induction motors

- Induction sealing

- Induction welding

- Inductive charging

- Inductors

- Magnetic flow meters

- Mechanically powered flashlight

- Pickups

- Rowland ring

- Transcranial magnetic stimulation

- Transformers

- Wireless energy transfer

Electrical Generator

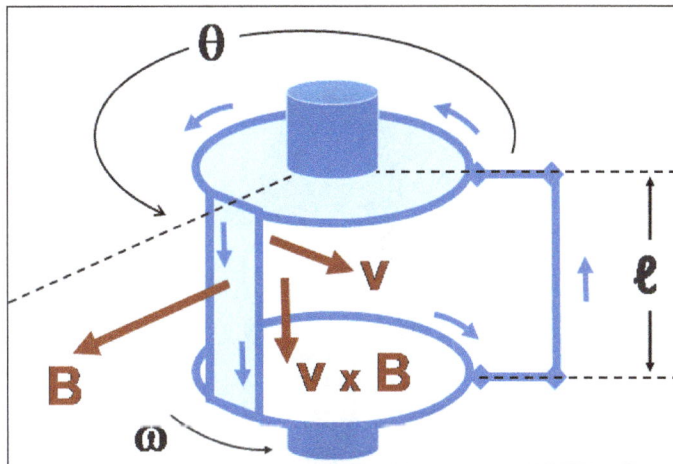

Rectangular wire loop rotating at angular velocity ω in radially outward pointing magnetic field B of fixed magnitude. The circuit is completed by brushes making sliding contact with top and bottom discs, which have conducting rims. This is a simplified version of the drum generator.

The EMF generated by Faraday's law of induction due to relative movement of a circuit and a magnetic field is the phenomenon underlying electrical generators. When a permanent magnet is moved relative to a conductor, or vice versa, an electromotive force is created. If the wire is connected through an electrical load, current will flow, and thus electrical energy is generated, converting the mechanical energy of motion to electrical energy. For example, the *drum generator* is based upon the figure to the bottom-right. A different implementation of this idea is the Faraday's disc, shown in simplified form on the right.

In the Faraday's disc example, the disc is rotated in a uniform magnetic field perpendicular to the disc, causing a current to flow in the radial arm due to the Lorentz force. Mechanical work is necessary to drive this current. When the generated current flows through the conducting rim, a magnetic field is generated by this current through Ampère's circuital law (labelled "induced B" in the figure). The rim thus becomes an electromagnet that resists rotation of the disc (an example of Lenz's law). On the far side of the figure, the return current flows from the rotating arm through the far side of the rim to the bottom brush. The B-field induced by this return current opposes the applied B-field, tending to *decrease* the flux through that side of the circuit, opposing the *increase* in flux due to rotation. On the near side of the figure, the return current flows from the rotating arm through the near side of the rim to the bottom brush. The induced B-field *increases* the flux on this side of the circuit, opposing the *decrease* in flux due to r the rotation. The energy required to keep the disc moving, despite this reactive force, is exactly equal to the electrical energy generated (plus energy wasted due to friction, Joule heating, and other inefficiencies). This behavior is common to all generators converting mechanical energy to electrical energy.

Electrical Transformer

When the electric current in a loop of wire changes, the changing current creates a changing magnetic field. A second wire in reach of this magnetic field will experience this change in magnetic field as a change in its coupled magnetic flux, $d \Phi_B / d t$. Therefore, an electromotive force is set up in the second loop called the induced EMF or transformer EMF. If the two ends of this loop are connected through an electrical load, current will flow.

Current Clamp

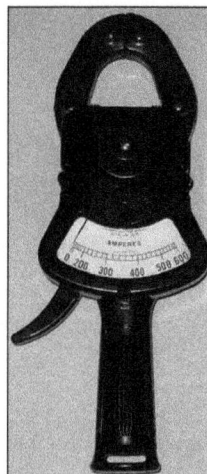

A current clamp.

A current clamp is a type of transformer with a split core which can be spread apart and clipped onto a wire or coil to either measure the current in it or, in reverse, to induce a voltage. Unlike conventional instruments the clamp does not make electrical contact with the conductor or require it to be disconnected during attachment of the clamp.

Magnetic Flow Meter

Faraday's law is used for measuring the flow of electrically conductive liquids and slurries. Such instruments are called magnetic flow meters. The induced voltage ε generated in the magnetic field B due to a conductive liquid moving at velocity v is thus given by:

$$\mathcal{E} = -B\ell v,$$

where ℓ is the distance between electrodes in the magnetic flow meter.

Eddy Currents

Conductors (of finite dimensions) moving through a uniform magnetic field, or stationary within a changing magnetic field, will have currents induced within them. These induced eddy currents can be undesirable, since they dissipate energy in the resistance of the conductor. There are a number of methods employed to control these undesirable inductive effects:

- Electromagnets in electric motors, generators, and transformers do not use solid metal, but instead use thin sheets of metal plate, called *laminations*. These thin plates reduce the parasitic eddy currents.

- Inductive coils in electronics typically use magnetic cores to minimize parasitic current flow. They are a mixture of metal powder plus a resin binder that can hold any shape. The binder prevents parasitic current flow through the powdered metal.

Electromagnet Laminations

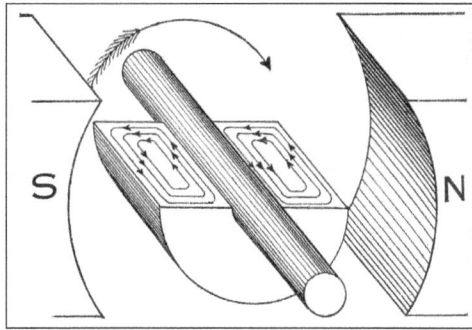

Eddy currents occur when a solid metallic mass is rotated in a magnetic field, because the outer portion of the metal cuts more lines of force than the inner portion, hence the induced electromotive force not being uniform, tends to set up currents between the points of greatest and least potential. Eddy currents consume a considerable amount of energy and often cause a harmful rise in temperature.

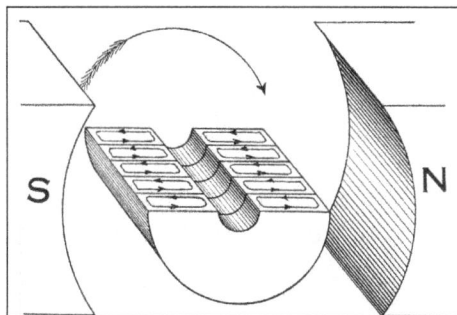

Only five laminations or plates are shown in this example, so as to show the subdivision of the eddy currents. In practical use, the number of laminations or punchings ranges from 40 to 66 per inch, and brings the eddy current loss down to about one percent. While the plates can be separated by insulation, the voltage is so low that the natural rust/oxide coating of the plates is enough to prevent current flow across the laminations.

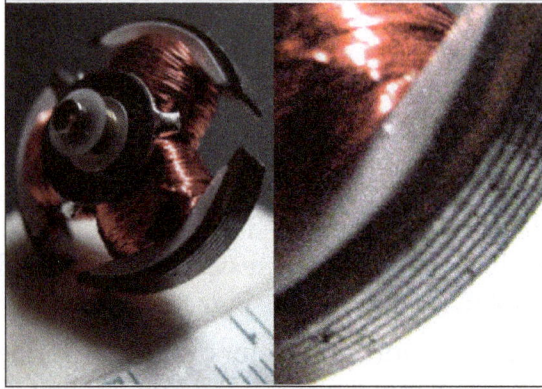

This is a rotor approximately 20mm in diameter from a DC motor used in a CD player. Note the laminations of the electromagnet pole pieces, used to limit parasitic inductive losses.

Parasitic Induction within Conductors

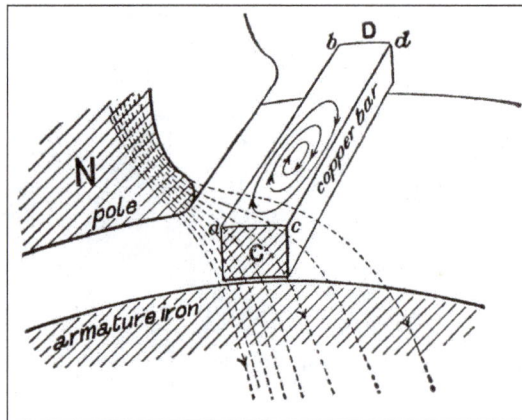

In this illustration, a solid copper bar conductor on a rotating armature is just passing under the tip of the pole piece N of the field magnet. Note the uneven distribution of the lines of force across the copper bar. The magnetic field is more concentrated and thus stronger on the left edge of the copper bar (a,b) while the field is weaker on the right edge (c,d). Since the two edges of the bar move with the same velocity, this difference in field strength across the bar creates whorls or current eddies within the copper bar.

High current power-frequency devices, such as electric motors, generators and transformers, use multiple small conductors in parallel to break up the eddy flows that can form within large solid conductors. The same principle is applied to transformers used at higher than power frequency, for example, those used in switch-mode power supplies and the intermediate frequency coupling transformers of radio receivers.

Displacement Current

In electromagnetism, displacement current density is the quantity $\partial D/\partial t$ appearing in Maxwell's equations that is defined in terms of the rate of change of D, the electric displacement field. Displacement current density has the same units as electric current density, and it is a source of the magnetic field just as actual current is. However, it is not an electric current of moving charges, but a time-varying electric field. In physical materials (as opposed to vacuum), there is also a contribution from the slight motion of charges bound in atoms, called dielectric polarization.

The idea was conceived by James Clerk Maxwell in his 1861 paper *On Physical Lines of Force, Part III* in connection with the displacement of electric particles in a dielectric medium. Maxwell added displacement current to the electric current term in Ampère's Circuital Law. In his 1865 paper A Dynamical Theory of the Electromagnetic Field Maxwell used this amended version of Ampère's Circuital Law to derive the electromagnetic wave equation. This derivation is now generally accepted as a historical landmark in physics by virtue of uniting electricity, magnetism and optics into one single unified theory. The displacement current term is now seen as a crucial addition that completed Maxwell's equations and is necessary to explain many phenomena, most particularly the existence of electromagnetic waves.

The electric displacement field is defined as:

$$D = \varepsilon_0 E + P.$$

where:

> ε_o is the permittivity of free space
>
> E is the electric field intensity
>
> P is the polarization of the medium

Differentiating this equation with respect to time defines the *displacement current density*, which therefore has two components in a dielectric:

$$J_D = \varepsilon_0 \frac{\partial E}{\partial t} + \frac{\partial P}{\partial t}.$$

The first term on the right hand side is present in material media and in free space. It doesn't necessarily come from any actual movement of charge, but it does have an associated magnetic field, just as a current does due to charge motion. Some authors apply the name *displacement current* to the first term by itself.

The second term on the right hand side, called polarization current density, comes from the change in polarization of the individual molecules of the dielectric material. Polarization results when, under the influence of an applied electric field, the charges in molecules have moved from a position of exact cancellation. The positive and negative charges in molecules separate, causing an increase in the state of polarization P. A changing state of polarization corresponds to charge movement and so is equivalent to a current, hence the term "polarization current".

Thus, $I_D = \iint\limits_S J_D \cdot dS = \iint\limits_S \dfrac{\partial D}{\partial t} \cdot dS = \dfrac{\partial}{\partial t} \iint\limits_S D \cdot dS = \dfrac{\partial \Phi_D}{\partial t}$

This polarization is the displacement current as it was originally conceived by Maxwell. Maxwell made no special treatment of the vacuum, treating it as a material medium. For Maxwell, the effect of P was simply to change the relative permittivity ε_r in the relation $D = \varepsilon_r \varepsilon_o E$.

The modern justification of displacement current is explained below.

Isotropic Dielectric Case

In the case of a very simple dielectric material the constitutive relation holds:

$\qquad D = \varepsilon E$,

where the permittivity $\varepsilon = \varepsilon_o \varepsilon_r$,

- ε_r is the relative permittivity of the dielectric and

- ε_o is the electric constant.

In this equation the use of ε accounts for the polarization of the dielectric.

The scalar value of displacement current may also be expressed in terms of electric flux:

$\qquad I_D = \varepsilon \dfrac{\partial \Phi_E}{\partial t}$.

The forms in terms of ε are correct only for linear isotropic materials. More generally ε may be replaced by a tensor, may depend upon the electric field itself, and may exhibit frequency dependence (dispersion).

For a linear isotropic dielectric, the polarization P is given by:

$\qquad P = \varepsilon_0 \chi_e E = \varepsilon_0 (\varepsilon_r - 1) E$

where χ_e is known as the electric susceptibility of the dielectric. Note that:

$\qquad \varepsilon = \varepsilon_r \varepsilon_0 = (1 + \chi_e) \varepsilon_0$.

Necessity

Some implications of the displacement current follow, which agree with experimental observation, and with the requirements of logical consistency for the theory of electromagnetism.

Generalizing Ampère's Circuital Law

Current in Capacitors

An example illustrating the need for the displacement current arises in connection with capacitors with no medium between the plates. Consider the charging capacitor in the figure. The capacitor is in a circuit that causes equal and opposite charges to appear on the left plate and the right plate, charging the capacitor and increasing the electric field between its plates. No actual charge is transported through the vacuum between its plates. Nonetheless, a magnetic field exists between the plates as though a current were present there as well. One explanation is that a *displacement current* I_D "flows" in the vacuum, and this current produces the magnetic field in the region between the plates according to Ampère's law:

$$\oint_C B \cdot d\ell = \mu_0 I_D \, .$$

where

- $\oint_C$ is the closed line integral around some closed curve C.

- B is the magnetic field measured in teslas.

- $\cdot$ is the vector dot product.

- $d\ell$ is an infinitesimal line element along the curve C, that is, a vector with magnitude equal to the length element of C, and direction given by the tangent to the curve C.

- μ_0 is the magnetic constant, also called the permeability of free space.

- I_D is the net displacement current that passes through a small surface bounded by the curve C.

The magnetic field between the plates is the same as that outside the plates, so the displacement current must be the same as the conduction current in the wires, that is,

$$I_D = I,$$

which extends the notion of current beyond a mere transport of charge.

Next, this displacement current is related to the charging of the capacitor. Consider the current in the imaginary cylindrical surface shown surrounding the left plate. A current, say I, passes outward through the left surface L of the cylinder, but no conduction current (no transport of real charges) crosses the right surface R. Notice that the electric field between the plates E increases as the capacitor charges. That is, in a manner described by Gauss's law, assuming no dielectric between the plates:

$$Q(t) = \varepsilon_0 \oint_S dS \cdot \mathbf{E}(t),$$

where S refers to the imaginary cylindrical surface. Assuming a parallel plate capacitor with uniform electric field, and neglecting fringing effects around the edges of the plates, differentiation provides:

$$\frac{dQ}{dt} = -I = -\varepsilon_0 \oint_S dS \cdot \frac{\partial E}{\partial t} \approx -S \, \varepsilon_0 \frac{\partial E}{\partial t},$$

where the sign is negative because charge leaves this plate (the charge is decreasing), and where S is the area of the surface R. The electric field at surface L is nearly zero because the field due to charge on the left-hand plate is nearly cancelled by the equal but opposite charge on the right-hand plate. Under the assumption of a uniform electric field distribution inside the capacitor, the displacement current density J_D is found by dividing by the area of the surface:

$$J_D = \frac{I_D}{S} = \frac{I}{S} = \varepsilon_0 \frac{\partial E}{\partial t} = \frac{\partial D}{\partial t},$$

where I is the current leaving the cylindrical surface (which must equal I_D) and J_D is the flow of charge per unit area into the cylindrical surface through the face R.

Combining these results, the magnetic field is found using the integral form of Ampère's law with an arbitrary choice of contour provided the displacement current density term is added to the conduction current density (the Ampère-Maxwell equation):

$$\oint_{\partial S} B \cdot d\ell = \mu_0 \int_S \left(J + \int_0 \frac{\partial E}{\partial t} \right) \cdot dS$$

This equation says that the integral of the magnetic field B around a loop ∂S is equal to the integrated current J through any surface spanning the loop, plus the displacement current term $\varepsilon_o \partial E / \partial t$ through the surface.

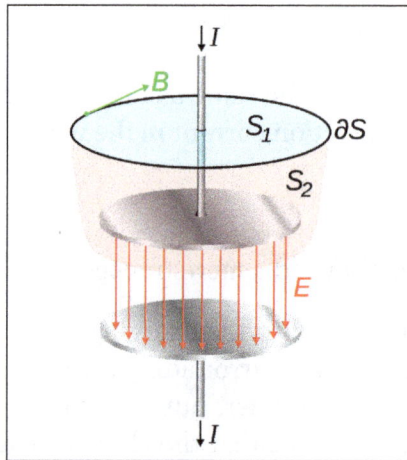

Example showing two surfaces S_1 and S_2 that share the same bounding contour ∂S. However, S_1 is pierced by conduction current, while S_2 is pierced by displacement current.

As depicted in the figure to the right, the current crossing surface S_1 is entirely conduction current. Applying the Ampère-Maxwell equation to surface S_1 yields:

$$B = \frac{\mu_0 I}{2\pi r}$$

However, the current crossing surface S_2 is entirely displacement current. Applying this law to surface S_2, which is bounded by exactly the same curve ∂S, but lies between the plates, produces:

$$B = \frac{\mu_0 I_D}{2\pi r}$$

Any surface S_1 that intersects the wire has current I passing through it so Ampère's law gives the correct magnetic field. However a second surface S_2 bounded by the same loop δS could be drawn passing between the capacitor plates, therefore having no current passing through it. Without the displacement current term Ampere's law would give zero magnetic field for this surface. Therefore without the displacement current term Ampere's law gives inconsistent results, the magnetic field would depend on the surface chosen for integration. Thus the displacement current term $\varepsilon_o \, \partial E / \partial t$ is necessary as a second source term which gives the correct magnetic field when the surface of integration passes between the capacitor plates. Because the current is increasing the charge on the capacitor's plates, the electric field between the plates is increasing, and the rate of change of electric field gives the correct value for the field B found above.

Mathematical Formulation

In a more mathematical vein, the same results can be obtained from the underlying differential equations. Consider for simplicity a non-magnetic medium where the relative magnetic permeability is unity, and the complication of magnetization current (bound current) is absent, so that $M=0$ and $J=J_f$. The current leaving a volume must equal the rate of decrease of charge in a volume. In differential form this continuity equation becomes:

$$\nabla \cdot J_f = -\frac{\partial \rho_f}{\partial t},$$

where the left side is the divergence of the free current density and the right side is the rate of decrease of the free charge density. However, Ampère's law in its original form states:

$$\nabla \times B = \mu_0 J_f,$$

which implies that the divergence of the current term vanishes, contradicting the continuity equation. (Vanishing of the divergence is a result of the mathematical identity that states the divergence of a curl is always zero.) This conflict is removed by addition of the displacement current, as then:

$$\nabla \times B = \mu_0 \left(J + \varepsilon_0 \frac{\partial E}{\partial t} \right) = \mu_0 \left(J_f + \frac{\partial D}{\partial t} \right),$$

And

$$\nabla \cdot (\nabla \times B) = 0 = \mu_0 \left(\nabla \cdot J_f + \frac{\partial}{\partial t} \nabla \cdot D \right),$$

which is in agreement with the continuity equation because of Gauss's law:

$$\nabla \cdot D = \rho_f .$$

Wave Propagation

The added displacement current also leads to wave propagation by taking the curl of the equation for magnetic field.

$$J_D = \epsilon_0 \frac{\partial E}{\partial t}$$

Substituting this form for J into Ampère's law, and assuming there is no bound or free current density contributing to J :

$$\nabla \times B = \mu_0 J_D \,,$$

with the result:

$$\nabla \times (\nabla \times B) = \mu_0 \, \epsilon_0 \frac{\partial}{\partial t} \nabla \times E \,.$$

However,

$$\nabla \times E = -\frac{\partial}{\partial t} B \,,$$

leading to the wave equation:

$$-\nabla \times (\nabla \times B) = \nabla^2 B = \mu_0 \, \epsilon_0 \frac{\partial^2}{\partial t^2} B = \frac{1}{c^2} \frac{\partial^2}{\partial t^2} B \,,$$

where use is made of the vector identity that holds for any vector field V(r, t):

$$\nabla \times (\nabla \times V) = \nabla (\nabla \cdot V) - \nabla^2 V \,,$$

and the fact that the divergence of the magnetic field is zero. An identical wave equation can be found for the electric field by taking the curl:

$$\nabla \times (\nabla \times E) = -\frac{\partial}{\partial t} \nabla \times B = -\mu_0 \frac{\partial}{\partial t} \left(J + \epsilon_0 \frac{\partial}{\partial t} E \right) .$$

If J, P and ρ are zero, the result is:

$$\nabla^2 E = \mu_0 \, \epsilon_0 \frac{\partial^2}{\partial t^2} E = \frac{1}{c^2} \frac{\partial^2}{\partial t^2} E \,.$$

The electric field can be expressed in the general form:

$$E = -\nabla \varphi - \frac{\partial A}{\partial t} \,,$$

where φ is the electric potential (which can be chosen to satisfy Poisson's equation) and A is a vector potential (i.e. magnetic vector potential, not to be confused with Surface area, as A is denoted elsewhere). The $\nabla \varphi$ component on the right hand side is the Gauss's law component, and this is the component that is relevant to the conservation of charge argument above. The second term on the right-hand side is the one relevant to the electromagnetic wave equation, because it is the term that contributes to the *curl* of E. Because of the vector identity that says the *curl* of a *gradient* is zero, $\nabla\varphi$ does not contribute to $\nabla\times E$.

Maxwell's Equations

Maxwell's equations are a set of coupled partial differential equations that, together with the Lorentz force law, form the foundation of classical electromagnetism, classical optics, and electric circuits. The equations provide a mathematical model for electric, optical, and radio technologies, such as power generation, electric motors, wireless communication, lenses, radar etc. Maxwell's equations describe how electric and magnetic fields are generated by charges, currents, and changes of the fields. An important consequence of the equations is that they demonstrate how fluctuating electric and magnetic fields propagate at a constant speed (c) in a vacuum. Known as electromagnetic radiation, these waves may occur at various wavelengths to produce a spectrum of light from radio waves to γ-rays. The equations are named after the physicist and mathematician James Clerk Maxwell, who between 1861 and 1862 published an early form of the equations that included the Lorentz force law. Maxwell first used the equations to propose that light is an electromagnetic phenomenon.

The equations have two major variants. The microscopic Maxwell equations have universal applicability but are unwieldy for common calculations. They relate the electric and magnetic fields to total charge and total current, including the complicated charges and currents in materials at the atomic scale. The "macroscopic" Maxwell equations define two new auxiliary fields that describe the large-scale behaviour of matter without having to consider atomic scale charges and quantum phenomena like spins. However, their use requires experimentally determined parameters for a phenomenological description of the electromagnetic response of materials.

The term "Maxwell's equations" is often also used for equivalent alternative formulations. Versions of Maxwell's equations based on the electric and magnetic potentials are preferred for explicitly solving the equations as a boundary value problem, analytical mechanics, or for use in quantum mechanics. The covariant formulation (on spacetime rather than space and time separately) makes the compatibility of Maxwell's equations with special relativity manifest. Maxwell's equations in curved spacetime, commonly used in high energy and gravitational physics, are compatible with general relativity. In fact, Einstein developed special and general relativity to accommodate the invariant speed of light, a consequence of Maxwell's equations, with the principle that only relative movement has physical consequences.

Since the mid-20th century, it has been understood that Maxwell's equations are not exact, but a classical limit of the fundamental theory of quantum electrodynamics.

Formulation in Terms of Electric and Magnetic Fields (Microscopic or in Vacuum Version)

In the electric and magnetic field formulation there are four equations that determine the fields for given charge and current distribution. A separate law of nature, the Lorentz force law, describes how, conversely, the electric and magnetic field act on charged particles and currents. A version of this law was included in the original equations by Maxwell but, by convention, is included no longer. The vector calculus formalism below, due to Oliver Heaviside, has become standard. It is manifestly rotation invariant, and therefore mathematically much more transparent than Maxwell's original 20 equations in x,y,z components. The relativistic formulations are even more symmetric and manifestly Lorentz invariant.

The differential and integral equations formulations are mathematically equivalent and are both useful. The integral formulation relates fields within a region of space to fields on the boundary and can often be used to simplify and directly calculate fields from symmetric distributions of charges and currents. On the other hand, the differential equations are purely *local* and are a more natural starting point for calculating the fields in more complicated (less symmetric) situations, for example using finite element analysis.

Key to the Notation

Symbols in bold represent vector quantities, and symbols in *italics* represent scalar quantities, unless otherwise indicated. The equations introduce the electric field, E, a vector field, and the magnetic field, B, a pseudovector field, each generally having a time and location dependence. The sources are

- The total electric charge density (total charge per unit volume), ρ, and

- The total electric current density (total current per unit area), J.

The universal constants appearing in the equations (the first two ones explicitly only in the SI units formulation) are:

- The permittivity of free space, ε_0, and

- The permeability of free space, μ_0, and

- The speed of light, $c = \dfrac{1}{\sqrt{\varepsilon_0 \mu_0}}$

Differential Equations

In the differential equations,

- The nabla symbol, ∇, denotes the three-dimensional gradient operator, del,

- The $\nabla \cdot$ symbol (pronounced "del dot") denotes the divergence operator,

- The $\nabla \times$ symbol (pronounced "del cross") denotes the curl operator.

Integral Equations

In the integral equations,

- Ω is any fixed volume with closed boundary surface $\partial\Omega$,

- Σ is any fixed surface with closed boundary curve $\partial\Sigma$,

Here a *fixed* volume or surface means that it does not change over time. The equations are correct, complete, and a little easier to interpret with time-independent surfaces. For example, since the surface is time-independent, we can bring the differentiation under the integral sign in Faraday's law:

$$\frac{d}{dt}\iint_\Sigma B \cdot dS = \iint_\Sigma \frac{\partial B}{\partial t} \cdot dS ,$$

Maxwell's equations can be formulated with possibly time-dependent surfaces and volumes by using the differential version and using Gauss and Stokes formula appropriately.

- $\oiint_{\partial\Omega}$ is a surface integral over the boundary surface $\partial\Omega$, with the loop indicating the surface is closed

- $\iiint_\Omega$ is a volume integral over the volume Ω,

- $\oint_{\partial\Sigma}$ is a line integral around the boundary curve $\partial\Sigma$, with the loop indicating the curve is closed.

- $\iint_\Sigma$ is a surface integral over the surface Σ,

- The total electric charge Q enclosed in Ω is the volume integral over Ω of the charge density ρ:

$$Q = \iiint_\Omega \rho \, dV ,$$

where dV is the volume element.

- The net electric current I is the surface integral of the electric current density J passing through a fixed surface, Σ:

$$I = \iint_\Sigma J \cdot dS,$$

where dS denotes the differential vector element of surface area S, normal to surface Σ. (Vector area is sometimes denoted by A rather than S, but this conflicts with the notation for magnetic potential).

Formulation in SI Units Convention

Name	Integral equations	Differential equations
Gauss's law	$\oiint_{\partial\Omega} E \cdot dS = \frac{1}{\varepsilon_0} \iiint_\Omega \rho \, dV$	$\nabla \cdot E = \frac{\rho}{\varepsilon_0}$
Gauss's law for magnetism	$\oiint_{\partial\Omega} B \cdot dS = 0$	$\nabla \cdot B = 0$

Maxwell–Faraday equation (Faraday's law of induction)	$\oint_{\partial\Sigma} E \cdot dl = -\dfrac{d}{dt} \iint_{\Sigma} B \cdot dS$	$\nabla \times E = -\dfrac{\partial B}{\partial t}$
Ampère's circuital law (with Maxwell's addition)	$\oint_{\partial\Sigma} B \cdot dl = \mu_0 \left(\iint_{\Sigma} J \cdot dS + \varepsilon_0 \dfrac{d}{dt} \iint_{\Sigma} E \cdot dS \right)$	$\nabla \times B = \mu_0 \left(J + \varepsilon_0 \dfrac{\partial E}{\partial t} \right)$

Formulation in Gaussian Units Convention

The definitions of charge, electric field, and magnetic field can be altered to simplify theoretical calculation, by absorbing dimensioned factors of ε_0 and μ_0 into the units of calculation, by convention. With a corresponding change in convention for the Lorentz force law this yields the same physics, i.e. trajectories of charged particles, or work done by an electric motor. These definitions are often preferred in theoretical and high energy physics where it is natural to take the electric and magnetic field with the same units, to simplify the appearance of the electromagnetic tensor: the Lorentz covariant object unifying electric and magnetic field would then contain components with uniform unit and dimension. Such modified definitions are conventionally used with the Gaussian (CGS) units. Using these definitions and conventions, colloquially "in Gaussian units", the Maxwell equations become:

Name	Integral equations	Differential equations
Gauss's law	$\oiint_{\partial\Omega} E \cdot dS = 4\pi \iiint_{\Omega} \rho \, dV$	$\nabla \cdot E = 4\pi\rho$
Gauss's law for magnetism	$\oiint_{\partial\Omega} B \cdot dS = 0$	$\nabla \cdot B = 0$
Maxwell–Faraday equation (Faraday's law of induction)	$\oint_{\partial\Sigma} E \cdot d\ell = -\dfrac{1}{c} \dfrac{d}{dt} \iint_{\Sigma} B \cdot dS$	$\nabla \times E = -\dfrac{1}{c} \dfrac{\partial B}{\partial t}$
Ampère's circuital law (with Maxwell's addition)	$\oint_{\partial\Sigma} B \cdot d\ell = \dfrac{1}{c} \left(4\pi \iint_{\Sigma} J \cdot dS + \dfrac{d}{dt} \iint_{\Sigma} E \cdot dS \right)$	$\nabla \times B = \dfrac{1}{c} \left(4\pi J + \dfrac{\partial E}{\partial t} \right)$

The equations are particularly readable when length and time are measured in compatible units like seconds and lightseconds i.e. in units such that c = 1 unit of length/unit of time. Ever since 1983, metres and seconds are compatible except for historical legacy since *by definition* c = 299 792 458 m/s ($\approx$ 1.0 feet/nanosecond).

Relationship between Differential and Integral Formulations

The equivalence of the differential and integral formulations are a consequence of the Gauss divergence theorem and the Kelvin–Stokes theorem.

Flux and Divergence

According to the (purely mathematical) Gauss divergence theorem, the electric flux through the boundary surface $\partial\Omega$ can be rewritten as:

$$\oiint_{\partial\Omega} E \cdot dS = \iiint_{\Omega} \nabla \cdot E \, dV$$

The integral version of Gauss's equation can thus be rewritten as:

$$\iiint_\Omega \left(\nabla \cdot \mathbf{E} - \frac{\rho}{\epsilon_0} \right) dV = 0$$

Since Ω is arbitrary (e.g. an arbitrary small ball with arbitrary center), this is satisfied iff the integrand is zero. This is the differential equations formulation of Gauss equation up to a trivial rearrangement.

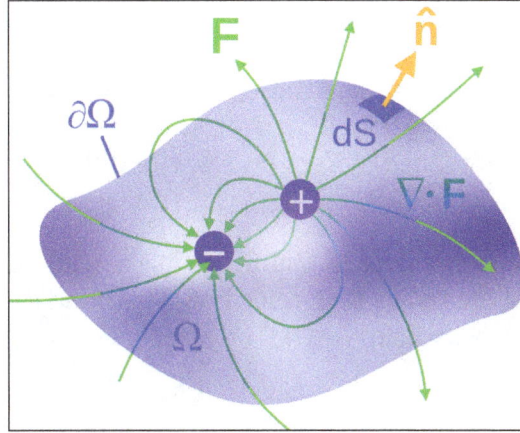

Volume Ω and its closed boundary $\partial\Omega$, containing (respectively enclosing) a source (+) and sink (−) of a vector field F. Here, F could be the E field with source electric charges, but *not* the B field, which has no magnetic charges as shown. The outward unit normal is n.

Similarly rewriting the magnetic flux in Gauss's law for magnetism in integral form gives

$$\oiint_{\partial\Omega} B \cdot dS = \iiint_\Omega \nabla \cdot B \, dV = 0.$$

which is satisfied for all Ω iff $\nabla \cdot B = 0$.

Circulation and Curl

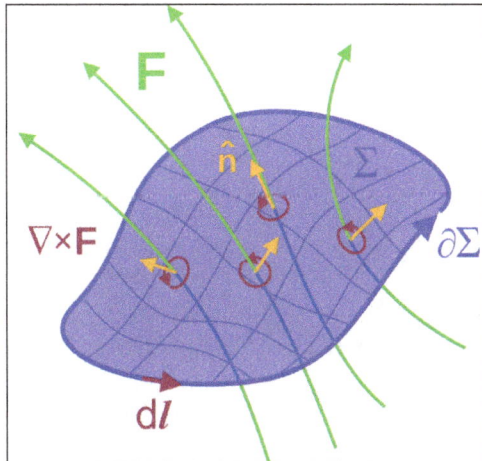

Surface Σ with closed boundary $\partial\Sigma$. F could be the E or B fields. Again, n is the unit normal. (The curl of a vector field doesn't literally look like the "circulations", this is a heuristic depiction.)

By the Kelvin–Stokes theorem we can rewrite the line integrals of the fields around the closed boundary curve $\partial\Sigma$ to an integral of the "circulation of the fields" (i.e. their curls) over a surface it bounds, i.e.

$$\int_{\partial\Sigma} B\cdot d\ell = \iint_{\Sigma} (\nabla\times B)\cdot dS,$$

Hence the modified Ampere law in integral form can be rewritten as,

$$\iint_{\Sigma} \left(\nabla\times B - \mu_0\left(J + \epsilon_0\frac{\partial E}{\partial t}\right)\right)\cdot dS = 0.$$

Since Σ can be chosen arbitrarily, e.g. as an arbitrary small, arbitrary oriented, and arbitrary centered disk, we conclude that the integrand is zero iff Ampere's modified law in differential equations form is satisfied. The equivalence of Faraday's law in differential and integral form follows likewise.

The line integrals and curls are analogous to quantities in classical fluid dynamics: the circulation of a fluid is the line integral of the fluid's flow velocity field around a closed loop, and the vorticity of the fluid is the curl of the velocity field.

Charge Conservation

The invariance of charge can be derived as a corollary of Maxwell's equations. The left-hand side of the modified Ampere's Law has zero divergence by the div–curl identity. Expanding the divergence of the right-hand side, interchanging derivatives, and applying Gauss's law gives:

$$0 = \nabla\cdot\nabla\times B = \mu_0\left(\nabla\cdot J + \varepsilon_0\frac{\partial}{\partial t}\nabla\cdot E\right) = \mu_0\left(\nabla\cdot J + \frac{\partial\rho}{\partial t}\right)$$

i.e.,

$$\frac{\partial\rho}{\partial t} + \nabla\cdot J = 0.$$

By the Gauss Divergence Theorem, this means the rate of change of charge in a fixed volume equals the net current flowing through the boundary:

$$\frac{d}{dt}Q_\Omega = \frac{d}{dt}\iiint_\Omega \rho dV = -\oiint_{\partial\Omega} J\cdot dS = -I_{\partial\Omega}.$$

In particular, in an isolated system the total charge is conserved.

Vacuum Equations, Electromagnetic Waves and Speed of Light

In a region with no charges ($\rho = 0$) and no currents ($J = 0$), such as in a vacuum, Maxwell's equations reduce to:

$$\nabla\cdot E = 0 \quad \nabla\times E = -\frac{\partial B}{\partial t},$$

$$\nabla \cdot B = 0 \quad \nabla \times B = \mu_0 \varepsilon_0 \frac{\partial E}{\partial t}.$$

Taking the curl ($\nabla \times$) of the curl equations, and using the curl of the curl identity we obtain:

$$\mu_0 \varepsilon_0 \frac{\partial^2 E}{\partial t^2} - \nabla^2 E = 0$$

$$\mu_0 \varepsilon_0 \frac{\partial^2 B}{\partial t^2} - \nabla^2 B = 0$$

The quantity $\mu_0 \varepsilon_0$ has the dimension of (time/length)². Defining $c = (\mu_0 \varepsilon_0)^{-1/2}$ the equations above have the form of the standard wave equations:

$$\frac{1}{c^2} \frac{\partial^2 E}{\partial t^2} - \nabla^2 E = 0$$

$$\frac{1}{c^2} \frac{\partial^2 B}{\partial t^2} - \nabla^2 B = 0$$

Already during Maxwell's lifetime, it was found that the known values for ε_0 and μ_0 give $c \approx 2.998 \times 10^8$ m/s , then already known to be the speed of light in free space. This led him to propose that light and radio waves were propagating electromagnetic waves, since amply confirmed. In the old SI system of units, the values of $\mu_0 = 4\pi \times 10^{-7}$ and $c = 299792458$ m/s are defined constants, (which means that by definition $\varepsilon_0 = 8854 \times 10^{-12}$ F/m) that define the ampere and the metre. In the new SI system, only c keeps its defined value, and the electron charge gets a defined value.

In materials with relative permittivity, ε_r, and relative permeability, μ_r, the phase velocity of light becomes:

$$v_p = \frac{1}{\sqrt{\mu_0 \mu_r \varepsilon_0 \varepsilon_r}}$$

which is usually less than c.

In addition, E and B are perpendicular to each other and to the direction of wave propagation, and are in phase with each other. A sinusoidal plane wave is one special solution of these equations. Maxwell's equations explain how these waves can physically propagate through space. The changing magnetic field creates a changing electric field through Faraday's law. In turn, that electric field creates a changing magnetic field through Maxwell's addition to Ampère's law. This perpetual cycle allows these waves, now known as electromagnetic radiation, to move through space at velocity c.

Macroscopic Formulation

The above equations are the "microscopic" version of Maxwell's equations, expressing the electric and the magnetic fields in terms of the (possibly atomic-level) charges and currents present. This is sometimes called the "general" form, but the macroscopic version below is equally general, the difference being one of bookkeeping.

The microscopic version is sometimes called "Maxwell's equations in a vacuum": this refers to the fact that the material medium is not built into the structure of the equations, but appears only in the charge and current terms. The microscopic version was introduced by Lorentz, who tried to use it to derive the macroscopic properties of bulk matter from its microscopic constituents.

"Maxwell's macroscopic equations", also known as Maxwell's equations in matter, are more similar to those that Maxwell introduced himself.

Name	Integral equations (SI convention)	Differential equations (SI convention)	Differential equations (Gaussian convention)
Gauss's law	$\oiint_{\partial\Omega} D \cdot dS = \iiint_\Omega \rho_f \, dV$	$\nabla \cdot D = \rho_f$	$\nabla \cdot D = 4\pi\rho_f$
Gauss's law for magnetism	$\oiint_{\partial\Omega} B \cdot dS = 0$	$\nabla \cdot B = 0$	$\nabla \cdot B = 0$
Maxwell–Faraday equation (Faraday's law of induction)	$\oint_{\partial\Sigma} E \cdot d\ell = -\dfrac{d}{dt} \iint_\Sigma B \cdot dS$	$\nabla \times E = -\dfrac{\partial B}{\partial t}$	$\nabla \times E = -\dfrac{1}{c}\dfrac{\partial B}{\partial t}$
Ampère's circuital law (with Maxwell's addition)	$\oint_{\partial\Sigma} H \cdot d\ell = \iint_\Sigma J_f \cdot dS + \dfrac{d}{dt} \iint_\Sigma D \cdot dS$	$\nabla \times H = J_f + \dfrac{\partial D}{\partial t}$	$\nabla \times H = \dfrac{1}{c}\left(4\pi J_f + \dfrac{\partial D}{\partial t}\right)$

In the "macroscopic" equations, the influence of bound charge Q_b and bound current I_b is incorporated into the displacement field D and the magnetizing field H, while the equations depend only on the free charges Q_f and free currents I_f. This reflects a splitting of the total electric charge Q and current I (and their densities ρ and J) into free and bound parts:

$$Q = Q_f + Q_b = \iiint_\Omega \left(\rho_f + \rho_b\right) dV = \iiint_\Omega \rho \, dV$$

$$I = I_f + I_b = \iint_\Sigma \left(J_f + J_b\right) \cdot dS = \iint_\Sigma J \cdot dS$$

The cost of this splitting is that the additional fields D and H need to be determined through phenomenological constituent equations relating these fields to the electric field E and the magnetic field B, together with the bound charge and current.

Bound Charge and Current

When an electric field is applied to a dielectric material its molecules respond by forming microscopic electric dipoles – their atomic nuclei move a tiny distance in the direction of the field, while their electrons move a tiny distance in the opposite direction. This produces a *macroscopic bound charge* in the material even though all of the charges involved are bound to individual molecules. For example, if every molecule responds the same, similar to that shown in the figure, these tiny movements of charge combine to produce a layer of positive bound charge on one side of the material and a layer of negative charge on the other side. The bound charge is most conveniently described in terms of the polarization P of the material, its dipole moment per unit volume. If P is uniform, a macroscopic separation of charge is produced only at the surfaces where P enters and leaves the material. For non-uniform P, a charge is also produced in the bulk.

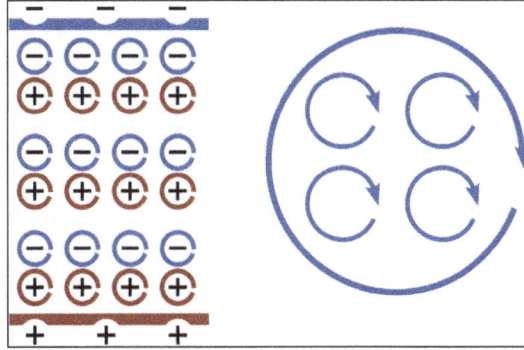

Left: A schematic view of how an assembly of microscopic dipoles produces opposite surface charges as shown at top and bottom. *Right:* How an assembly of microscopic current loops add together to produce a macroscopically circulating current loop. Inside the boundaries, the individual contributions tend to cancel, but at the boundaries no cancelation occurs.

Somewhat similarly, in all materials the constituent atoms exhibit magnetic moments that are intrinsically linked to the angular momentum of the components of the atoms, most notably their electrons. The connection to angular momentum suggests the picture of an assembly of microscopic current loops. Outside the material, an assembly of such microscopic current loops is not different from a macroscopic current circulating around the material's surface, despite the fact that no individual charge is traveling a large distance. These *bound currents* can be described using the magnetization M.

The very complicated and granular bound charges and bound currents, therefore, can be represented on the macroscopic scale in terms of P and M, which average these charges and currents on a sufficiently large scale so as not to see the granularity of individual atoms, but also sufficiently small that they vary with location in the material. As such, *Maxwell's macroscopic equations* ignore many details on a fine scale that can be unimportant to understanding matters on a gross scale by calculating fields that are averaged over some suitable volume.

Auxiliary Fields, Polarization and Magnetization

The *definitions* (not constitutive relations) of the auxiliary fields are:

$$D(r,t) = \varepsilon_0 E(r,t) + P(r,t)$$

$$H(r,t) = \frac{1}{\mu_0} B(r,t) - M(r,t)$$

where P is the polarization field and M is the magnetization field, which are defined in terms of microscopic bound charges and bound currents respectively. The macroscopic bound charge density ρ_b and bound current density J_b in terms of polarization P and magnetization M are then defined as,

$$\rho_b = -\nabla \cdot P$$

$$J_b = \nabla \times M + \frac{\partial P}{\partial t}$$

If we define the total, bound, and free charge and current density by,

$$\rho = \rho_b + \rho_f,$$

$$J = J_b + J_f,$$

and use the defining relations above to eliminate D, and H, the "macroscopic" Maxwell's equations reproduce the "microscopic" equations.

Constitutive Relations

In order to apply 'Maxwell's macroscopic equations', it is necessary to specify the relations between displacement field D and the electric field E, as well as the magnetizing field H and the magnetic field B. Equivalently, we have to specify the dependence of the polarization P (hence the bound charge) and the magnetization M (hence the bound current) on the applied electric and magnetic field. The equations specifying this response are called constitutive relations. For real-world materials, the constitutive relations are rarely simple, except approximately, and usually determined by experiment. For materials without polarization and magnetization, the constitutive relations are,

$$D = \varepsilon_0 E, \quad H = \frac{1}{\mu_0} B$$

where ε_0 is the permittivity of free space and μ_0 the permeability of free space. Since there is no bound charge, the total and the free charge and current are equal.

An alternative viewpoint on the microscopic equations is that they are the macroscopic equations *together* with the statement that vacuum behaves like a perfect linear "material" without additional polarization and magnetization. More generally, for linear materials the constitutive relations are,

$$D = \varepsilon E, \quad H = \frac{1}{\mu} B$$

where ε is the permittivity and μ the permeability of the material. For the displacement field D the linear approximation is usually excellent because for all but the most extreme electric fields or temperatures obtainable in the laboratory (high power pulsed lasers) the interatomic electric fields of materials of the order of 10^{11} V/m are much higher than the external field. For the magnetizing field H, however, the linear approximation can break down in common materials like iron leading to phenomena like hysteresis. Even the linear case can have various complications, however.

- For homogeneous materials, ε and μ are constant throughout the material, while for inhomogeneous materials they depend on location within the material (and perhaps time).

- For isotropic materials, ε and μ are scalars, while for anisotropic materials (e.g. due to crystal structure) they are tensors.

- Materials are generally dispersive, so ε and μ depend on the frequency of any incident EM waves.

Even more generally, in the case of non-linear materials, D and P are not necessarily proportional to E, similarly H or M is not necessarily proportional to B. In general D and H depend on both E and B, on location and time, and possibly other physical quantities.

In applications one also has to describe how the free currents and charge density behave in terms of E and B possibly coupled to other physical quantities like pressure, and the mass, number density, and velocity of charge-carrying particles. E.g., the original equations given by Maxwell included Ohm's law in the form,

$$J_f = \sigma E.$$

Alternative Formulations

Following is a summary of some of the numerous other mathematical formalisms to write the microscopic Maxwell's equations, with the columns separating the two homogeneous Maxwell equations from the two inhomogeneous ones involving charge and current. Each formulation has versions directly in terms of the electric and magnetic fields, and indirectly in terms of the electrical potential φ and the vector potential A. Potentials were introduced as a convenient way to solve the homogeneous equations, but it was thought that all observable physics was contained in the electric and magnetic fields (or relativistically, the Faraday tensor). The potentials play a central role in quantum mechanics, however, and act quantum mechanically with observable consequences even when the electric and magnetic fields vanish (Aharonov–Bohm effect). Each table describes one formalism. SI units are used throughout.

Vector Calculus

Formulation	Homogeneous equations	Inhomogeneous equations
Fields 3D Euclidean space + time	$\nabla \cdot B = 0$ $\nabla \times E + \dfrac{\partial B}{\partial t} = 0$	$\nabla \cdot E = \dfrac{\rho}{\varepsilon_0}$ $\nabla \times B - \dfrac{1}{c^2}\dfrac{\partial E}{\partial t} = \mu_0 J$
Potentials (any gauge) 3D Euclidean space + time	$B = \nabla \times A$ $E = -\nabla \varphi - \dfrac{\partial A}{\partial t}$	$-\nabla^2 \varphi - \dfrac{\partial}{\partial t}\left(\nabla \cdot A\right) = \dfrac{\rho}{\varepsilon_0}$ $\left(-\nabla^2 + \dfrac{1}{c^2}\dfrac{\partial^2}{\partial t^2}\right)A + \nabla\left(\nabla \cdot A + \dfrac{1}{c^2}\dfrac{\partial \varphi}{\partial t}\right) = \mu_0 J$
Potentials (Lorenz gauge) 3D Euclidean space + time	$B = \nabla \times A$ $E = -\nabla \varphi - \dfrac{\partial A}{\partial t}$ $\nabla \cdot A = -\dfrac{1}{c^2}\dfrac{\partial \varphi}{\partial t}$	$\left(-\nabla^2 + \dfrac{1}{c^2}\dfrac{\partial^2}{\partial t^2}\right)\varphi = \dfrac{\rho}{\varepsilon_0}$ $\left(-\nabla^2 + \dfrac{1}{c^2}\dfrac{\partial^2}{\partial t^2}\right)A = \mu_0 J$

Tensor Calculus

Formulation	Homogeneous equations	Inhomogeneous equations
Fields space + time spatial metric independent of time	$\partial_{[i}B_{jk]} =$ $\nabla_{[i}B_{jk]} = 0$ $\partial_{[i}E_{j]} + \dfrac{\partial B_{ij}}{\partial t} =$ $\nabla_{[i}E_{j]} + \dfrac{\partial B_{ij}}{\partial t} = 0$	$\dfrac{1}{\sqrt{h}}\partial_i\sqrt{h}E^i =$ $\nabla_i E^i = \dfrac{\rho}{\epsilon_0}$ $-\dfrac{1}{\sqrt{h}}\partial_i\sqrt{h}B^{ij} - \dfrac{1}{c^2}\dfrac{\partial}{\partial t}E^j =$ $-\nabla_i B^{ij} - \dfrac{1}{c^2}\dfrac{\partial E^j}{\partial t} = \mu_0 J^j$
Potentials space (with topological restrictions) + time spatial metric independent of time	$B_{ij} = \partial_{[i}A_{j]}$ $= \nabla_{[i}A_{j]}$ $E_i = -\dfrac{\partial A_i}{\partial t} - \partial_i\phi$ $= -\dfrac{\partial A_i}{\partial t} - \nabla_i\phi$	$-\dfrac{1}{\sqrt{h}}\partial_i\sqrt{h}\left(\partial^i\phi + \dfrac{\partial A^i}{\partial t}\right) =$ $-\nabla_i\nabla^i\phi - \dfrac{\partial}{\partial t}\nabla_i A^i = \dfrac{\rho}{\epsilon_0}$ $-\dfrac{1}{\sqrt{h}}\partial_i\sqrt{h}h^{im}h^{jn}\partial_{[m}A_{n]} + \dfrac{1}{c^2}\dfrac{\partial}{\partial t}\left(\dfrac{\partial A^j}{\partial t} + \partial^j\phi\right) =$ $-\nabla_i\nabla^i A^j + \dfrac{1}{c^2}\dfrac{\partial^2 A^j}{\partial t^2} + R^j_i A^i + \nabla^j\left(\nabla_i A^i + \dfrac{1}{c^2}\dfrac{\partial\phi}{\partial t}\right) = \mu_0 J^j$
Potentials (Lorenz gauge) space (with topological restrictions) + time spatial metric independent of time	$B_{ij} = \partial_{[i}A_{j]}$ $= \nabla_{[i}A_{j]}$ $E_i = -\dfrac{\partial A_i}{\partial t} - \partial_i\phi$ $= -\dfrac{\partial A_i}{\partial t} - \nabla_i\phi$ $\nabla_i A^i = -\dfrac{1}{c^2}\dfrac{\partial\phi}{\partial t}$	$-\nabla_i\nabla^i\phi + \dfrac{1}{c^2}\dfrac{\partial^2\phi}{\partial t^2} = \dfrac{\rho}{\epsilon_0}$ $-\nabla_i\nabla^i A^j + \dfrac{1}{c^2}\dfrac{\partial^2 A^j}{\partial t^2} + R^j_i A^i = \mu_0 J^j$

Differential Forms

Formulation	Homogeneous equations	Inhomogeneous equations
Fields Any space + time	$dB = 0$ $dE + \dfrac{\partial B}{\partial t} = 0$	$d*E = \dfrac{\rho}{\epsilon_0}$ $d*B - \dfrac{1}{c^2}\dfrac{\partial *E}{\partial t} = \mu_0 J$
Potentials (any gauge) Any space (with topological restrictions) + time	$B = dA$ $E = -d\phi - \dfrac{\partial A}{\partial t}$	$-d*\left(d\phi + \dfrac{\partial A}{\partial t}\right) = \dfrac{\rho}{\epsilon_0}$ $d*dA + \dfrac{1}{c^2}\dfrac{\partial}{\partial t}*\left(d\phi + \dfrac{\partial A}{\partial t}\right) = \mu_0 J$
Potential (Lorenz Gauge) Any space (with topological restrictions) + time spatial metric independent of time	$B = dA$ $E = -d\phi - \dfrac{\partial A}{\partial t}$ $d*A = -*\dfrac{1}{c^2}\dfrac{\partial\phi}{\partial t}$	$*\left(-\Delta\phi + \dfrac{1}{c^2}\dfrac{\partial^2}{\partial t^2}\phi\right) = \dfrac{\rho}{\epsilon_0}$ $*\left(-\Delta A + \dfrac{1}{c^2}\dfrac{\partial^2 A}{\partial^2 t}\right) = \mu_0 J$

Relativistic Formulations

The Maxwell equations can also be formulated on a spacetime-like Minkowski space where space and time are treated on equal footing. The direct spacetime formulations make manifest that the Maxwell equations are relativistically invariant. Because of this symmetry electric and magnetic field are treated on equal footing and are recognised as components of the Faraday tensor. This reduces the four Maxwell equations to two, which simplifies the equations, although we can no longer use the familiar vector formulation. In fact the Maxwell equations in the space + time formulation are not Galileo invariant and have Lorentz invariance as a hidden symmetry. This was a major source of inspiration for the development of relativity theory. To repeat: the space + time formulation is not a non-relativistic approximation and it describes the same physics by simply renaming variables. For this reason the relativistic invariant equations are usually called the Maxwell equations as well. Each table describes one formalism.

Formulation	Homogeneous equations	Inhomogeneous equations
Fields Minkowski space	$\partial_{[\alpha}F_{\beta\gamma]} = 0$	$\partial_\alpha F^{\alpha\beta} = \mu_0 J^\beta$
Potentials (any gauge) Minkowski space	$F_{\alpha\beta} = 2\partial_{[\alpha}A_{\beta]}$	$2\partial_\alpha\partial^{[\alpha}A^{\beta]} = \mu_0 J^\beta$
Potentials (Lorenz gauge) Minkowski space	$F_{\alpha\beta} = 2\partial_{[\alpha}A_{\beta]}$ $\partial_\alpha A^\alpha = 0$	$\partial_\alpha\partial^\alpha A^\beta = \mu_0 J^\beta$
Fields Any spacetime	$\partial_{[\alpha}F_{\beta\gamma]} =$ $\nabla_{[\alpha}F_{\beta\gamma]} = 0$	$\dfrac{1}{\sqrt{-g}}\partial_\alpha(\sqrt{-g}F^{\alpha\beta}) =$ $\nabla_\alpha F^{\alpha\beta} = \mu_0 J^\beta$
Potentials (any gauge) Any spacetime (with topological restrictions)	$F_{\alpha\beta} = 2\partial_{[\alpha}A_{\beta]}$ $= 2\nabla_{[\alpha}A_{\beta]}$	$\dfrac{2}{\sqrt{-g}}\partial_\alpha(\sqrt{-g}g^{\alpha\mu}g^{\beta\nu}\partial_{[\mu}A_{\nu]}) =$ $2\nabla_\alpha(\nabla^{[\alpha}A^{\beta]}) = \mu_0 J^\beta$
Potentials (Lorenz gauge) Any spacetime (with topological restrictions)	$F_{\alpha\beta} = 2\partial_{[\alpha}A_{\beta]}$ $= 2\nabla_{[\alpha}A_{\beta]}$ $\nabla_\alpha A^\alpha = 0$	$\nabla_\alpha\nabla^\alpha A^\beta - R^\beta{}_\alpha A^\alpha = \mu_0 J^\beta$
Fields Any spacetime	$\mathrm{d}F = 0$	$\mathrm{d}\star F = \mu_0 J$
Potentials (any gauge) Any spacetime (with topological restrictions)	$F = \mathrm{d}A$	$\mathrm{d}\star\mathrm{d}A = \mu_0 J$
Potentials (Lorenz gauge) Any spacetime (with topological restrictions)	$F = \mathrm{d}A$ $\mathrm{d}\star A = 0$	$\star\Box A = \mu_0 J$

- In the tensor calculus formulation, the electromagnetic tensor $F_{\alpha\beta}$ is an antisymmetric covariant order 2 tensor; the four-potential, A_α, is a covariant vector; the current, J^α, is a vector; the square brackets, [ ], denote antisymmetrization of indices; ∂_α is the derivative with respect to the coordinate, x^α. In Minkowski space coordinates are chosen with respect to an inertial frame; $(x^\alpha) = (ct,x,y,z)$, so that the metric tensor used to raise and lower indices is

$\eta_{\alpha\beta} = \text{diag}(1,-1,-1,-1)$. The d'Alembert operator on Minkowski space is $\nabla = \partial_a\partial^a$ as in the vector formulation. In general spacetimes, the coordinate system x^a is arbitrary, the covariant derivative ∇_a, the Ricci tensor, $R_{\alpha\beta}$ and raising and lowering of indices are defined by the Lorentzian metric, $g_{\alpha\beta}$ and the d'Alembert operator is defined as $\nabla = \nabla_a\nabla^a$. The topological restriction is that the second real cohomology group of the space vanishes. This is violated for Minkowski space with a line removed, which can model a (flat) spacetime with a point-like monopole on the complement of the line.

- In the differential form formulation on arbitrary space times, $F = F_{\alpha\beta}dx^\alpha \wedge dx^\beta$ is the electromagnetic tensor considered as a 2-form, $A = A_a dx^\alpha$ is the potential 1-form, J is the current 3-form, d is the exterior derivative, and $\star$ is the Hodge star on forms defined (up to its an orientation, i.e. its sign) by the Lorentzian metric of spacetime. In the special case of 2-forms such as F, the Hodge star $\star$ depends on the metric tensor only for its local scale. This means that, as formulated, the differential form field equations are conformally invariant, but the Lorenz gauge condition breaks conformal invariance. The operator $\square = (-\star d \star d - d \star d \star)$ is the d'Alembert–Laplace–Beltrami operator on 1-forms on an arbitrary Lorentzian spacetime. The topological condition is again that the second real cohomology group is trivial. By the isomorphism with the second de Rham cohomology this condition means that every closed 2-form is exact.

Other formalisms include the geometric algebra formulation and a matrix representation of Maxwell's equations. Historically, a quaternionic formulation was used.

Maxwell's equations are partial differential equations that relate the electric and magnetic fields to each other and to the electric charges and currents. Often, the charges and currents are themselves dependent on the electric and magnetic fields via the Lorentz force equation and the constitutive relations. These all form a set of coupled partial differential equations which are often very difficult to solve: the solutions encompass all the diverse phenomena of classical electromagnetism. Some general remarks follow.

As for any differential equation, boundary conditions and initial conditions are necessary for a unique solution. For example, even with no charges and no currents anywhere in spacetime, there are the obvious solutions for which E and B are zero or constant, but there are also non-trivial solutions corresponding to electromagnetic waves. In some cases, Maxwell's equations are solved over the whole of space, and boundary conditions are given as asymptotic limits at infinity. In other cases, Maxwell's equations are solved in a finite region of space, with appropriate conditions on the boundary of that region, for example an artificial absorbing boundary representing the rest of the universe, or periodic boundary conditions, or walls that isolate a small region from the outside world (as with a waveguide or cavity resonator).

Jefimenko's equations (or the closely related Liénard–Wiechert potentials) are the explicit solution to Maxwell's equations for the electric and magnetic fields created by any given distribution of charges and currents. It assumes specific initial conditions to obtain the so-called "retarded solution", where the only fields present are the ones created by the charges. However, Jefimenko's equations are unhelpful in situations when the charges and currents are themselves affected by the fields they create.

Numerical methods for differential equations can be used to compute approximate solutions of Maxwell's equations when exact solutions are impossible. These include the finite element method and finite-difference time-domain method.

Overdetermination of Maxwell's Equations

Maxwell's equations *seem* overdetermined, in that they involve six unknowns (the three components of E and B) but eight equations (one for each of the two Gauss's laws, three vector components each for Faraday's and Ampere's laws). (The currents and charges are not unknowns, being freely specifiable subject to charge conservation.) This is related to a certain limited kind of redundancy in Maxwell's equations: It can be proven that any system satisfying Faraday's law and Ampere's law *automatically* also satisfies the two Gauss's laws, as long as the system's initial condition does. This explanation was first introduced by Julius Adams Stratton in 1941. Although it is possible to simply ignore the two Gauss's laws in a numerical algorithm (apart from the initial conditions), the imperfect precision of the calculations can lead to ever-increasing violations of those laws. By introducing dummy variables characterizing these violations, the four equations become not overdetermined after all. The resulting formulation can lead to more accurate algorithms that take all four laws into account.

Both identities $\nabla \cdot \nabla \times B \equiv 0, \nabla \cdot \nabla \times E \equiv 0$, which reduce eight equations to six independent ones, are the true reason of overdetermination.

Maxwell's Equations as the Classical Limit of QED

Maxwell's equations and the Lorentz force law (along with the rest of classical electromagnetism) are extraordinarily successful at explaining and predicting a variety of phenomena; however they are not exact, but a classical limit of quantum electrodynamics (QED).

Some observed electromagnetic phenomena are incompatible with Maxwell's equations. These include photon–photon scattering and many other phenomena related to photons or virtual photons, "nonclassical light" and quantum entanglement of electromagnetic fields. E.g. quantum cryptography cannot be described by Maxwell theory, not even approximately. The approximate nature of Maxwell's equations becomes more and more apparent when going into the extremely strong field regime or to extremely small distances.

Finally, Maxwell's equations cannot explain any phenomenon involving individual photons interacting with quantum matter, such as the photoelectric effect, Planck's law, the Duane–Hunt law, and single-photon light detectors. However, many such phenomena may be approximated using a halfway theory of quantum matter coupled to a classical electromagnetic field, either as external field or with the expected value of the charge current and density on the right hand side of Maxwell's equations.

Variations

Popular variations on the Maxwell equations as a classical theory of electromagnetic fields are relatively scarce because the standard equations have stood the test of time remarkably well.

Magnetic Monopoles

Maxwell's equations posit that there is electric charge, but no magnetic charge (also called magnetic monopoles), in the universe. Indeed, magnetic charge has never been observed, despite extensive searches, and may not exist. If they did exist, both Gauss's law for magnetism and Faraday's law would need to be modified, and the resulting four equations would be fully symmetric under the interchange of electric and magnetic fields.

Electromagnetic Field

An electromagnetic field (also EMF or EM field) is a magnetic field produced by moving electrically charged objects. It affects the behavior of non-comoving charged objects at any distance of the field. The electromagnetic field extends indefinitely throughout space and describes the electromagnetic interaction. It is one of the four fundamental forces of nature (the others are gravitation, weak interaction and strong interaction).

The field can be viewed as the combination of an electric field and a magnetic field. The electric field is produced by stationary charges, and the magnetic field by moving charges (currents); these two are often described as the sources of the field. The way in which charges and currents interact with the electromagnetic field is described by Maxwell's equations and the Lorentz force law. The force created by the electric field is much stronger than the force created by the magnetic field.

From a classical perspective in the history of electromagnetism, the electromagnetic field can be regarded as a smooth, continuous field, propagated in a wavelike manner; whereas from the perspective of quantum field theory, the field is seen as quantized, being composed of individual particles.

Structure

The electromagnetic field may be viewed in two distinct ways: a continuous structure or a discrete structure.

Continuous Structure

Classically, electric and magnetic fields are thought of as being produced by smooth motions of charged objects. For example, oscillating charges produce variations in electric and magnetic fields that may be viewed in a 'smooth', continuous, wavelike fashion. In this case, energy is viewed as being transferred continuously through the electromagnetic field between any two locations. For instance, the metal atoms in a radio transmitter appear to transfer energy continuously. This view is useful to a certain extent (radiation of low frequency), but problems are found at high frequencies.

Discrete Structure

The electromagnetic field may be thought of in a more 'coarse' way. Experiments reveal that in some circumstances electromagnetic energy transfer is better described as being carried in the form of packets called quanta (in this case, photons) with a fixed frequency. Planck's relation links the photon energy E of a photon to its frequency f through the equation:

$$E = hf$$

where h is Planck's constant, and f is the frequency of the photon . Although modern quantum optics tells us that there also is a semi-classical explanation of the photoelectric effect—the emission of electrons from metallic surfaces subjected to electromagnetic radiation—the photon was historically (although not strictly necessarily) used to explain certain observations. It is found that increasing the intensity of the incident radiation (so long as one remains in the linear regime) increases only the number of electrons ejected, and has almost no effect on the energy distribution of their ejection. Only the frequency of the radiation is relevant to the energy of the ejected electrons.

This quantum picture of the electromagnetic field (which treats it as analogous to harmonic oscillators) has proven very successful, giving rise to quantum electrodynamics, a quantum field theory describing the interaction of electromagnetic radiation with charged matter. It also gives rise to quantum optics, which is different from quantum electrodynamics in that the matter itself is modelled using quantum mechanics rather than quantum field theory.

Dynamics

In the past, electrically charged objects were thought to produce two different, unrelated types of field associated with their charge property. An electric field is produced when the charge is stationary with respect to an observer measuring the properties of the charge, and a magnetic field as well as an electric field is produced when the charge moves, creating an electric current with respect to this observer. Over time, it was realized that the electric and magnetic fields are better thought of as two parts of a greater whole — the electromagnetic field. Until 1820, when the Danish physicist H. C. Ørsted showed the effect of electric current on a compass needle, electricity and magnetism had been viewed as unrelated phenomena. In 1831, Michael Faraday made the seminal observation that time-varying magnetic fields could induce electric currents and then, in 1864, James Clerk Maxwell published his famous paper *A Dynamical Theory of the Electromagnetic Field*.

Once this electromagnetic field has been produced from a given charge distribution, other charged or magnetised objects in this field may experience a force. If these other charges and currents are comparable in size to the sources producing the above electromagnetic field, then a new net electromagnetic field will be produced. Thus, the electromagnetic field may be viewed as a dynamic entity that causes other charges and currents to move, and which is also affected by them. These interactions are described by Maxwell's equations and the Lorentz force law.

Feedback Loop

The behavior of the electromagnetic field can be divided into four different parts of a loop:

- The electric and magnetic fields are generated by moving electric charges.

- The electric and magnetic fields interact with each other.

- The electric and magnetic fields produce forces on electric charges.

- The electric charges move in space.

A common misunderstanding is that (a) the quanta of the fields act in the same manner as (b) the charged particles, such as electrons, that generate the fields. In our everyday world, electrons travel slowly through conductors with a drift velocity of a fraction of a centimeter (or inch) per second and through a vacuum tube at speeds of around 1 thousand km/s, but fields propagate at the speed of light, approximately 300 thousand kilometers (or 186 thousand miles) a second. The speed ratio between charged particles in a conductor and field quanta is on the order of one to a million. Maxwell's equations relate (a) the presence and movement of charged particles with (b) the generation of fields. Those fields can then affect the force on, and can then move other slowly moving charged particles. Charged particles can move at relativistic speeds nearing field propagation speeds, but, as Einstein showed, this requires enormous field energies, which are not present in our everyday experiences with electricity, magnetism, matter, and time and space.

The feedback loop can be summarized in a list, including phenomena belonging to each part of the loop:

- Charged particles generate electric and magnetic fields.

- The fields interact with each other:

    ◦ Changing electric field acts like a current, generating 'vortex' of magnetic field.

    ◦ Faraday induction: changing magnetic field induces (negative) vortex of electric field.

    ◦ Lenz's law: negative feedback loop between electric and magnetic fields.

- Fields act upon particles:

    ◦ Lorentz force: force due to electromagnetic field:

        ▪ electric force: same direction as electric field.

        ▪ Magnetic force: perpendicular both to magnetic field and to velocity of charge.

- Particles move:

    ◦ Current is movement of particles.

- Particles generate more electric and magnetic fields; cycle repeats.

Mathematical Description

There are different mathematical ways of representing the electromagnetic field. The first one views the electric and magnetic fields as three-dimensional vector fields. These vector fields each have a value defined at every point of space and time and are thus often regarded as functions of the space and time coordinates. As such, they are often written as E(x, y, z, t) (electric field) and B(x, y, z, t) (magnetic field).

If only the electric field (E) is non-zero, and is constant in time, the field is said to be an electrostatic field. Similarly, if only the magnetic field (B) is non-zero and is constant in time, the field is said to be a magnetostatic field. However, if either the electric or magnetic field has a time-dependence, then both fields must be considered together as a coupled electromagnetic field using Maxwell's equations.

With the advent of special relativity, physical laws became susceptible to the formalism of tensors. Maxwell's equations can be written in tensor form, generally viewed by physicists as a more elegant means of expressing physical laws.

The behaviour of electric and magnetic fields, whether in cases of electrostatics, magnetostatics, or electrodynamics (electromagnetic fields), is governed by Maxwell's equations. In the vector field formalism, these are:

$$\nabla \cdot E = \frac{\rho}{\varepsilon_0} \text{ (Gauss's law)}$$

$$\nabla \cdot B = 0 \text{ (Gauss's law for magnetism)}$$

$$\nabla \times E = -\frac{\partial B}{\partial t} \text{ (Faraday's law)}$$

$$\nabla \times B = \mu_0 J + \mu_0 \varepsilon_0 \frac{\partial E}{\partial t} \text{ (Maxwell–Ampère law)}$$

where ρ is the charge density, which can (and often does) depend on time and position, ϵ_0 is the permittivity of free space, μ_0 is the permeability of free space, and J is the current density vector, also a function of time and position. The units used above are the standard SI units. Inside a linear material, Maxwell's equations change by switching the permeability and permittivity of free space with the permeability and permittivity of the linear material in question. Inside other materials which possess more complex responses to electromagnetic fields, these terms are often represented by complex numbers, or tensors. The Lorentz force law governs the interaction of the electromagnetic field with charged matter.

When a field travels across to different media, the properties of the field change according to the various boundary conditions. These equations are derived from Maxwell's equations. The tangential components of the electric and magnetic fields as they relate on the boundary of two media are as follows:

$$E_1 = E_2$$
$$H_1 = H_2 \left(\text{current-free}\right)$$
$$D_1 = D_2 \left(\text{charge-free}\right)$$
$$B_1 = B_2$$

The angle of refraction of an electric field between media is related to the permittivity (ε) of each medium:

$$\frac{\tan \theta_1}{\tan \theta_2} = \frac{\varepsilon_{r2}}{\varepsilon_{r1}}$$

The angle of refraction of a magnetic field between media is related to the permeability (μ) of each medium:

$$\frac{\tan \theta_1}{\tan \theta_2} = \frac{\mu_{r2}}{\mu_{r1}}$$

Properties of the Field

Reciprocal Behavior of Electric and Magnetic Fields

The two Maxwell equations, Faraday's Law and the Ampère-Maxwell Law, illustrate a very practical feature of the electromagnetic field. Faraday's Law may be stated roughly as 'a changing magnetic field creates an electric field'. This is the principle behind the electric generator.

Ampere's Law roughly states that 'a changing electric field creates a magnetic field'. Thus, this law can be applied to generate a magnetic field and run an electric motor.

Behavior of the Fields in the Absence of Charges or Currents

Maxwell's equations take the form of an electromagnetic wave in a volume of space not containing charges or currents (free space) – that is, where ρ and J are zero. Under these conditions, the electric and magnetic fields satisfy the electromagnetic wave equation:

$$\left(\nabla^2 - \frac{1}{c^2} \frac{\partial^2}{\partial t^2} \right) E = 0$$

$$\left(\nabla^2 - \frac{1}{c^2} \frac{\partial^2}{\partial t^2} \right) B = 0$$

James Clerk Maxwell was the first to obtain this relationship by his completion of Maxwell's equations with the addition of a displacement current term to Ampere's circuital law.

Applications

Static E and M Fields and Static EM Fields

When an EM field is not varying in time, it may be seen as a purely electrical field or a purely magnetic field, or a mixture of both. However the general case of a static EM field with both electric and magnetic components present, is the case that appears to most observers. Observers who see only an electric or magnetic field component of a static EM field, have the other (electric or magnetic) component suppressed, due to the special case of the immobile state of the charges that produce the EM field in that case. In such cases the other component becomes manifest in other observer frames.

A consequence of this, is that any case that seems to consist of a "pure" static electric or magnetic field, can be converted to an EM field, with both E and M components present, by simply moving the observer into a frame of reference which is moving with regard to the frame in which only the "pure" electric or magnetic field appears. That is, a pure static electric field will show the familiar magnetic field associated with a current, in any frame of reference where the charge moves. Likewise, any new motion of a charge in a region that seemed previously to contain only a magnetic field, will show that the space now contains an electric field as well, which will be found to produces an additional Lorentz force upon the moving charge.

Thus, electrostatics, as well as magnetism and magnetostatics, are now seen as studies of the static EM field when a particular frame has been selected to suppress the other type of field, and since an EM field with both electric and magnetic will appear in any other frame, these "simpler" effects are merely the observer's.

Time-varying EM Fields in Maxwell's Equations

An EM field that varies in time has two "causes" in Maxwell's equations. One is charges and currents (so-called "sources"), and the other cause for an E or M field is a change in the other type of field (this last cause also appears in "free space" very far from currents and charges).

An electromagnetic field very far from currents and charges (sources) is called electromagnetic radiation (EMR) since it radiates from the charges and currents in the source, and has no "feedback" effect on them, and is also not affected directly by them in the present time (rather, it is indirectly produced by a sequences of changes in fields radiating out from them in the past). EMR consists of the radiations in the electromagnetic spectrum, including radio waves, microwave, infrared, visible light, ultraviolet light, X-rays, and gamma rays.

A notable application of visible light is that this type of energy from the Sun powers all life on Earth that either makes or uses oxygen.

A changing electromagnetic field which is physically close to currents and charges will have a dipole characteristic that is dominated by either a changing electric dipole, or a changing magnetic dipole. This type of dipole field near sources is called an electromagnetic *near-field*.

Changing electric dipole fields, as such, are used commercially as near-fields mainly as a source of dielectric heating. Otherwise, they appear parasitically around conductors which absorb EMR, and around antennas which have the purpose of generating EMR at greater distances.

Changing magnetic dipole fields (i.e., magnetic near-fields) are used commercially for many types of magnetic induction devices. These include motors and electrical transformers at low frequencies, and devices such as metal detectors and MRI scanner coils at higher frequencies. Sometimes these high-frequency magnetic fields change at radio frequencies without being far-field waves and thus radio waves.

Other

- Electromagnetic field can be used to record data on static electricity.

- Old televisions can be traced with Electromagnetic fields.

Electromagnetic Radiation

In classical physics Electromagnetic radiation is the flow of energy at the universal speed of light through free space or through a material medium in the form of the electric and magnetic fields that make up electromagnetic waves such as radio waves, visible light, and gamma rays. In such a wave, time-varying electric and magnetic fields are mutually linked with each other at right angles and perpendicular to the direction of motion. An electromagnetic wave is characterized by its intensity and the frequency ν of the time variation of the electric and magnetic fields.

In terms of the modern quantum theory, electromagnetic radiation is the flow of photons (also called light quanta) through space. Photons are packets of energy hν that always move with the universal speed of light. The symbol h is Planck's constant, while the value of ν is the same as that of the frequency of the electromagnetic wave of classical theory. Photons having the same energy hν are all alike, and their number density corresponds to the intensity of the radiation. Electromagnetic radiation exhibits a multitude of phenomena as it interacts with charged particles in atoms, molecules, and larger objects of matter. These phenomena as well as the ways in which electromagnetic radiation is created and observed, the manner in which such radiation occurs in nature, and its technological uses depend on its frequency ν. The spectrum of frequencies of electromagnetic radiation extends from very low values over the range of radio waves, television waves, and microwaves to visible light and beyond to the substantially higher values of ultraviolet light, X-rays, and gamma rays.

Occurrence and Importance

Close to 0.01 percent of the mass/energy of the entire universe occurs in the form of electromagnetic radiation. All human life is immersed in it, and modern communications technology and medical services are particularly dependent on one or another of its forms. In fact, all living things on Earth depend on the electromagnetic radiation received from the Sun and on the transformation of solar energy by photosynthesis into plant life or by biosynthesis into zooplankton, the basic step in the food chain in oceans. The eyes of many animals, including those of humans, are adapted to be sensitive to and hence to see the most abundant part of the Sun's electromagnetic radiation—namely, light, which comprises the visible portion of its wide range of frequencies. Green plants also have high sensitivity to the maximum intensity of solar electromagnetic radiation, which is absorbed by a substance called chlorophyll that is essential for plant growth via photosynthesis.

Practically all the fuels that modern society uses—gas, oil, and coal—are stored forms of energy received from the Sun as electromagnetic radiation millions of years ago. Only the energy from nuclear reactors does not originate from the Sun.

Everyday life is pervaded by artificially made electromagnetic radiation: Food is heated in microwave ovens, airplanes are guided by radar waves, television sets receive electromagnetic waves transmitted by broadcasting stations, and infrared waves from heaters provide warmth. Infrared waves also are given off and received by automatic self-focusing cameras that electronically measure and set the correct distance to the object to be photographed. As soon as the Sun sets, incandescent or fluorescent lights are turned on to provide artificial illumination, and cities glow brightly with the colourful fluorescent and neon lamps of advertisement signs. Familiar too is ultraviolet radiation, which the eyes cannot see but whose effect is felt as pain from sunburn. Ultraviolet light represents a kind of electromagnetic radiation that can be harmful to life. Such is also true of X-rays, which are important in medicine as they allow physicians to observe the inner parts of the body but exposure to which should be kept to a minimum. Less familiar are gamma rays, which come from nuclear reactions and radioactive decay and are part of the harmful high-energy radiation of radioactive materials and nuclear weapons.

The Electromagnetic Spectrum

The brief account of familiar phenomena given above surveyed electromagnetic radiation from low frequencies of ν (radio waves) to exceedingly high values of ν (gamma rays). Going from the ν values of radio waves to those of visible light is like comparing the thickness of this page with the distance of Earth from the Sun, which represents an increase by a factor of a million billion. Similarly, going from the ν values of visible light to the very much larger ones of gamma rays represents another increase in frequency by a factor of a million billion. This extremely large range of ν values, called the electromagnetic spectrum, is shown in figure, together with the common names used for its various parts, or regions.

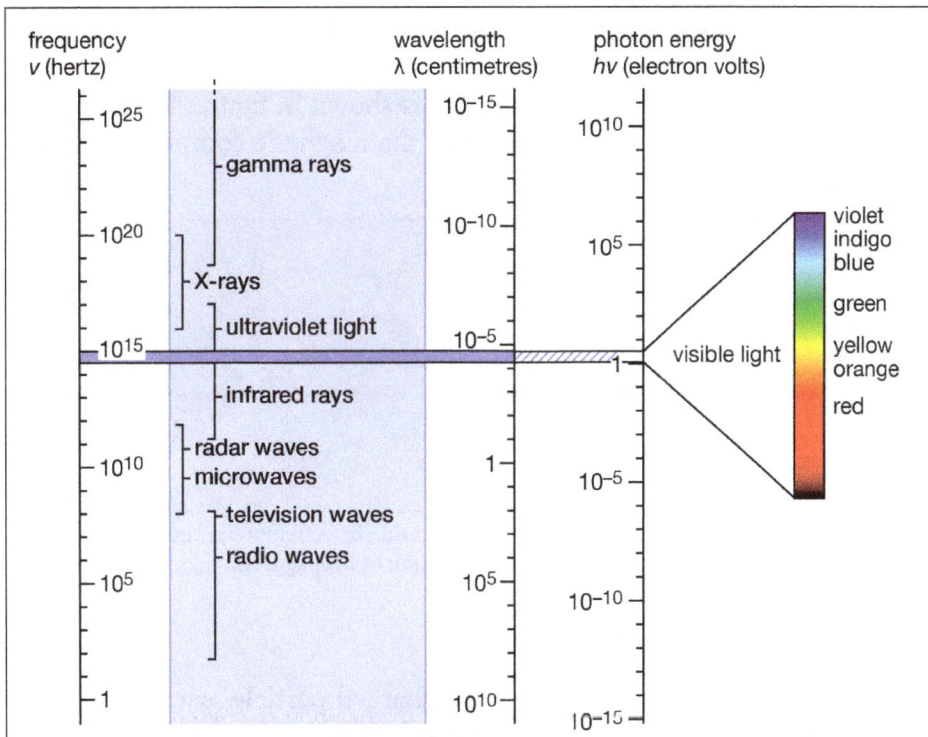

Electromagnetic spectrum. The small visible range (shaded) is shown enlarged at the right.

The number ν is shared by both the classical and the modern interpretation of electromagnetic radiation. In classical language, ν is the frequency of the temporal changes in an electromagnetic wave. The frequency of a wave is related to its speed c and wavelength λ in the following way. If 10 complete waves pass by in one second, one observes 10 wriggles, and one says that the frequency of such a wave is ν = 10 cycles per second (10 hertz [Hz]). If the wavelength of the wave is, say, λ = 3 cm, then it is clear that a wave train 30 cm long has passed in that one second to produce the 10 wriggles that were observed. Thus, the speed of the wave is 30 cm per second, and one notes that in general the speed is $c = \lambda v$. The speed of electromagnetic radiation of all kinds is the same universal constant that is defined to be exactly c = 299,792,458 metres per second (186,282 miles per second). The wavelengths of the classical electromagnetic waves in free space calculated from $c = \lambda v$ are also shown on the spectrum in figure, as is the energy hv of modern-day photons. Commonly used as the unit of energy is the electron volt (eV), which is the energy that can be given to an electron by a one-volt battery. It is clear that the range of wavelengths λ and of photon energies hv are equally as large as the spectrum of v values.

Because the wavelengths and energy quanta hv of electromagnetic radiation of the various parts of the spectrum are so different in magnitude, the sources of the radiations, the interactions with matter, and the detectors employed are correspondingly different. This is why the same electromagnetic radiation is called by different names in various regions of the spectrum.

In spite of these obvious differences of scale, all forms of electromagnetic radiation obey certain general rules that are well understood and that allow one to calculate with very high precision their properties and interactions with charged particles in atoms, molecules, and large objects. Electromagnetic radiation is, classically speaking, a wave of electric and magnetic fields propagating at the speed of light c through empty space. In this wave the electric and magnetic fields change their magnitude and direction each second. This rate of change is the frequency v measured in cycles per second—namely, in hertz. The electric and magnetic fields are always perpendicular to each other and at right angles to the direction of propagation, as shown in figure. There is as much energy carried by the electric component of the wave as by the magnetic component, and the energy is proportional to the square of the field strength.

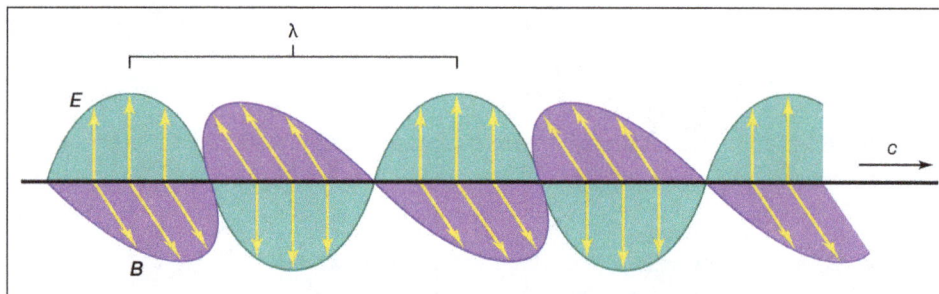

Radiation fields in which vectors {E vector} and {B vector} are perpendicular to each other and to the direction of propagation.

Generation of Electromagnetic Radiation

Electromagnetic radiation is produced whenever a charged particle, such as an electron, changes its velocity—i.e., whenever it is accelerated or decelerated. The energy of the electromagnetic radiation thus produced comes from the charged particle and is therefore lost by it. A common example of this phenomenon is the oscillating charge or current in a radio antenna. The antenna of a radio transmitter is part of an electric resonance circuit in which the charge is made to oscillate at a desired frequency. An electromagnetic wave so generated can be received by a similar antenna connected to an oscillating electric circuit in the tuner that is tuned to that same frequency. The electromagnetic wave in turn produces an oscillating motion of charge in the receiving antenna. In general, one can say that any system which emits electromagnetic radiation of a given frequency can absorb radiation of the same frequency.

Such human-made transmitters and receivers become smaller with decreasing wavelength of the electromagnetic wave and prove impractical in the millimetre range. At even shorter wavelengths down to the wavelengths of X-rays, which are one million times smaller, the oscillating charges arise from moving charges in molecules and atoms.

One may classify the generation of electromagnetic radiation into two categories: (1) systems or processes that produce radiation covering a broad continuous spectrum of frequencies and (2) those that emit (and absorb) radiation of discrete frequencies that are characteristic of particular systems. The Sun with its continuous spectrum is an example of the first, while a radio transmitter tuned to one frequency exemplifies the second category.

Continuous Spectra of Electromagnetic Radiation

Such spectra are emitted by any warm substance. Heat is the irregular motion of electrons, atoms, and molecules; the higher the temperature, the more rapid the motion. Since electrons are much lighter than atoms, irregular thermal motion produces irregular oscillatory charge motion, which reflects a continuous spectrum of frequencies. Each oscillation at a particular frequency can be considered a tiny "antenna" that emits and receives electromagnetic radiation. As a piece of iron is heated to increasingly high temperatures, it first glows red, then yellow, and finally white. In short, all the colours of the visible spectrum are represented. Even before the iron begins to glow red, one can feel the emission of infrared waves by the heat sensation on the skin. A white-hot piece of iron also emits ultraviolet radiation, which can be detected by a photographic film.

Not all materials heated to the same temperature emit the same amount and spectral distribution of electromagnetic waves. For example, a piece of glass heated next to iron looks nearly colourless, but it feels hotter to the skin (it emits more infrared rays) than does the iron. This observation illustrates the rule of reciprocity: a body radiates strongly at those frequencies that it is able to absorb, because for both processes it needs the tiny antennas of that range of frequencies. Glass is transparent in the visible range of light because it lacks possible electronic absorption at these particular frequencies. As a consequence, glass cannot glow red because it cannot absorb red. On the other hand, glass is a better emitter/absorber in the infrared than iron or any other metal that strongly reflects such lower-frequency electromagnetic waves. This selective emissivity and absorptivity is important for understanding the greenhouse effect and many other phenomena in nature. The tungsten filament of a lightbulb has a temperature of 2,500 K (4,040 °F) and emits large amounts of visible light but relatively little infrared because metals, have small emissivities in the infrared range. This is of course fortunate, since one wants light from a lightbulb but not much heat. The light emitted by a candle originates from very hot carbon soot particles in the flame, which strongly absorb and thus emit visible light. By contrast, the gas flame of a kitchen range is pale, even though it is hotter than a candle flame, because of the absence of soot. Light from the stars originates from the high temperature of the gases at their surface. A wide spectrum of radiation is emitted from the Sun's surface, the temperature of which is about 5,800 K. The radiation output is 60 million watts for every square metre of solar surface, which is equivalent to the amount produced by an average-size commercial power-generating station that can supply electric power for about 30,000 households.

The spectral composition of a heated body depends on the materials of which the body consists. That is not the case for an ideal radiator or absorber. Such an ideal object absorbs and thus emits radiation of all frequencies equally and fully. A radiator/absorber of this kind is called a blackbody, and its radiation spectrum is referred to as blackbody radiation, which depends on only one parameter, its temperature. Scientists devise and study such ideal objects because their properties can be known exactly. This information can then be used to determine and understand why real objects, such as a piece of iron or glass, a cloud, or a star, behave differently.

A good approximation of a blackbody is a piece of coal or, better yet, a cavity in a piece of coal that is visible through a small opening. There is one property of blackbody radiation which is familiar to everyone but which is actually quite mysterious. As the piece of coal is heated to higher and higher temperatures, one first observes a dull red glow, followed by a change in colour to bright red; as the temperature is increased further, the colour changes to yellow and finally to white. White is not itself a colour but rather the visual effect of the combination of all primary colours. The fact that white glow is observed at high temperatures means that the colour blue has been added to the ones observed at lower temperatures. This colour change with temperature is mysterious because one would expect, as the energy (or temperature) is increased, just more of the same and not something entirely different. For example, as one increases the power of a radio amplifier, one hears the music louder but not at a higher pitch.

The change in colour or frequency distribution of the electromagnetic radiation coming from heated bodies at different temperatures remained an enigma for centuries. The solution of this mystery by the German physicist Max Planck initiated the era of modern physics at the beginning of the 20th century. He explained the phenomenon by proposing that the tiny antennas in the heated body are quantized, meaning that they can emit electromagnetic radiation only in finite energy quanta of size hv. The universal constant h is called Planck's constant in his honour. For blue light hv = 3 eV, whereas hv = 1.8 eV for red light. Since high-frequency antennas of vibrating charges in solids have to emit larger energy quanta hv than lower-frequency antennas, they can only do so when the temperature, or the thermal atomic motion, becomes high enough. Hence, the average pitch, or peak frequency, of blackbody electromagnetic radiation increases with temperature.

The many tiny antennas in a heated chunk of material are, as noted above, to be identified with the accelerating and decelerating charges in the heat motion of the atoms of the material. There are other sources of continuous spectra of electromagnetic radiation that are not associated with heat but still come from accelerated or decelerated charges. X-rays are, for example, produced by abruptly stopping rapidly moving electrons. This deceleration of the charges produces bremsstrahlung ("braking radiation"). In an X-ray tube, electrons moving with an energy of E_{max} = 10,000 to 50,000 eV (10–50 keV) are made to strike a piece of metal. The electromagnetic radiation produced by this sudden deceleration of electrons is a continuous spectrum extending up to the maximum photon energy $hv = E_{max}$.

By far the brightest continuum spectra of electromagnetic radiation come from synchrotron radiation sources. These are not well known because they are predominantly used for research and sometimes for commercial and medical applications. Because any change in motion is an acceleration, circulating currents of electrons produce electromagnetic radiation. When these circulating electrons move at relativistic speeds (i.e., those approaching the speed of light), the brightness of the radiation increases enormously. This radiation was first observed at the General Electric Company in 1947 in an electron synchrotron (hence the name of this radiation), which is a type of particle accelerator that forces relativistic electrons into circular orbits by using powerful magnetic fields. The intensity of synchrotron radiation is further increased more than a thousandfold by wigglers and undulators that move the beam of relativistic electrons to and fro by means of other magnetic fields.

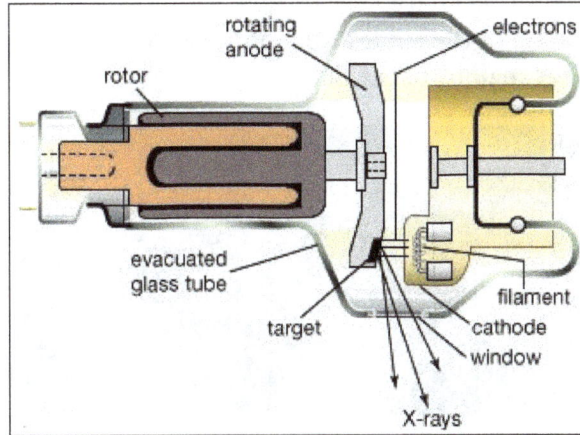

An X-ray tube. Electrons "boil" off the cathode when the filament is heated by a current. A high voltage between cathode and anode causes the electrons to accelerate toward the anode, which rotates to avoid overheating of the target. When the electrons strike the anode's target area, X-rays are emitted.

The conditions for generating bremsstrahlung as well as synchrotron radiation exist in nature in various forms. Acceleration and capture of charged particles by the gravitational field of a star, black hole, or galaxy is a source of energetic cosmic X-rays. Gamma rays are produced in other kinds of cosmic objects—namely, supernovae, neutron stars, and quasars.

Discrete-frequency Sources and Absorbers of Electromagnetic Radiation

These are commonly encountered in everyday life. Familiar examples of discrete-frequency electromagnetic radiation include the distinct colours of lamps filled with different fluorescent gases that are characteristic of advertisement signs, the colours of dyes and pigments, the bright yellow of sodium lamps, the blue-green hue of mercury lamps, and the specific colours of lasers.

Sources of electromagnetic radiation of specific frequency are typically atoms or molecules. Every atom or molecule can have certain discrete internal energies, which are called quantum states. An atom or molecule can therefore change its internal energy only by discrete amounts. By going from a higher to a lower energy state, a quantum hv of electromagnetic radiation is emitted of a magnitude that is precisely the energy difference between the higher and lower state. Absorption of a quantum hv takes the atom from a lower to a higher state if hv matches the energy difference. All like atoms are identical, but each chemical element of the periodic table has its own specific set of possible internal energies. Therefore, by measuring the characteristic and discrete electromagnetic radiation that is either emitted or absorbed by atoms or molecules, one can identify which kind of atom or molecule is giving off or absorbing the radiation. This provides a means of determining the chemical composition of substances. Since one cannot subject a piece of a distant star to conventional chemical analysis, studying the emission or absorption of starlight is the only way to determine the composition of stars or of interstellar gases and dust.

The Sun, for example, not only emits the continuous spectrum of radiation that originates from its hot surface but also emits discrete radiation quanta hv that are characteristic of its atomic composition. Many of the elements can be detected at the solar surface, but the most abundant is helium. This is so because helium is the end product of the nuclear fusion reaction that is the fundamental energy source of the Sun. This particular element was named helium because its existence was first discovered by its characteristic absorption energies in the Sun's spectrum. The helium of the cooler outer parts of the solar atmosphere absorbs the characteristic light frequencies from the lower and hotter regions of the Sun.

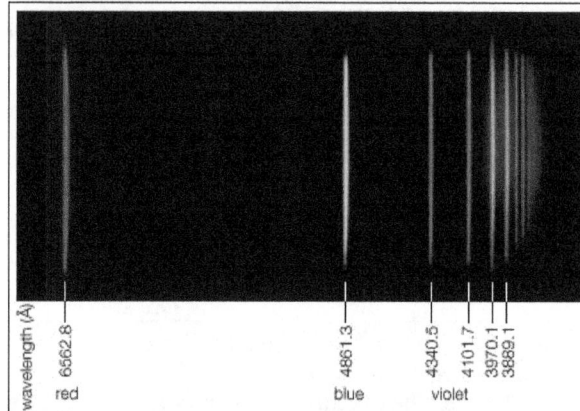

The Balmer series of atomic hydrogen. These lines are emitted when the electron in the hydrogen atom transitions from the $n = 3$ or greater orbital down to the $n = 2$ orbital. The wavelengths of these lines are given by $1/\lambda = R_H (1/4 - 1/n^2)$, where λ is the wavelength, R_H is the Rydberg constant, and n is the level of the original orbital.

The characteristic and discrete energies $h\nu$ found as emission and absorption of electromagnetic radiation by atoms and molecules extend to X-ray energies. As high-energy electrons strike the piece of metal in an X-ray tube, electrons are knocked out of the inner energy shell of the atoms. These vacancies are then filled by electrons from the second or third shell; emitted in the process are X-rays having $h\nu$ values that correspond to the energy differences of the shells. One therefore observes not only the continuous spectrum of the bremsstrahlung but also X-ray emissions of discrete energies $h\nu$ that are characteristic of the specific elemental composition of the metal struck by the energetic electrons in the X-ray tube.

The discrete electromagnetic radiation energies $h\nu$ emitted or absorbed by all substances reflect the discreteness of the internal energies of all material things. This means that window glass and water are transparent to visible light; they cannot absorb these visible light quanta because their internal energies are such that no energy difference between a higher and a lower internal state matches the energy $h\nu$ of visible light. Figure shows as an example the coefficient of absorption of water as a function of frequency ν of electromagnetic radiation. Above the scale of frequencies, the corresponding scales of photon energy $h\nu$ and wavelength λ are given. An absorption coefficient $\alpha = 10^{-4}$ cm^{-1} means that the intensity of electromagnetic radiation is only one-third its original value after passing through 100 metres of water. When $\alpha = 1$ cm^{-1}, only a layer 1 cm thick is needed to decrease the intensity to one-third its original value, and, for $\alpha = 10^3$ cm, a layer of water having the thickness of a thin sheet of paper is sufficient to attenuate electromagnetic radiation by that much. The transparency of water to visible light, marked by the vertical dashed lines, is a remarkable feature that is significant for life on Earth.

All things look so different and have different colours because of their different sets of internal discrete energies, which determine their interaction with electromagnetic radiation. The words looking and colours are associated with the human detectors of electromagnetic radiation, the eyes. Since there are instruments available for detecting electromagnetic radiation of any frequency, one can imagine that things "look" different at different energies of the spectrum because different materials have their own characteristic sets of discrete internal energies. Even the nuclei of atoms are composites of other elementary particles and thus can be excited to many discrete internal energy states. Since nuclear energies are much larger than atomic energies, the energy differences between internal energy states are substantially larger, and the corresponding electromagnetic radiation quanta hv emitted or absorbed when nuclei change their energies are even bigger than those of X-rays. Such quanta given off or absorbed by atomic nuclei are called gamma rays.

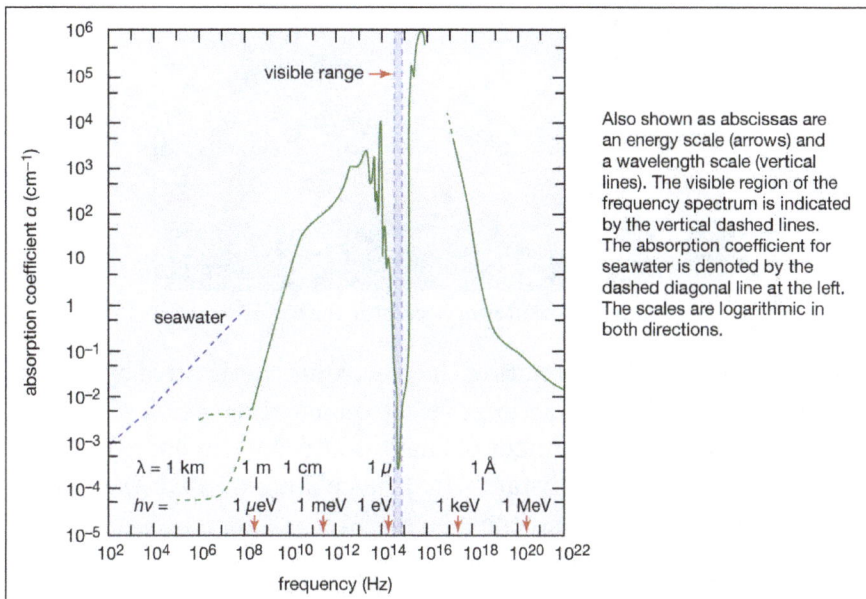

The absorption coefficient for liquid water as a function of frequency.

Properties and Behavior

Scattering, Reflection and Refraction

If a charged particle interacts with an electromagnetic wave, it experiences a force proportional to the strength of the electric field and thus is forced to change its motion in accordance with the frequency of the electric field wave. In doing so, it becomes a source of electromagnetic radiation of the same frequency. The energy for the work done in accelerating the charged particle and emitting this secondary radiation comes from and is lost by the primary wave. This process is called scattering.

Since the energy density of the electromagnetic radiation is proportional to the square of the electric field strength and the field strength is caused by acceleration of a charge, the energy radiated by such a charge oscillator increases with the square of the acceleration. On the other hand, the acceleration of an oscillator depends on the frequency of the back-and-forth oscillation. The acceleration increases with the square of the frequency. This leads to the important result that the electromagnetic energy radiated by an oscillator increases very rapidly—namely, with the square of the square or, as one says, with the fourth power of the frequency. Doubling the frequency thus produces an increase in radiated energy by a factor of 16.

This rapid increase in scattering with the frequency of electromagnetic radiation can be seen on any sunny day: it is the reason the sky is blue and the setting Sun is red. The higher-frequency blue light from the Sun is scattered much more by the atoms and molecules of Earth's atmosphere than is the lower-frequency red light. Hence, the light of the setting Sun, which passes through a thick layer of atmosphere, has much more red than yellow or blue light, while light scattered from the sky contains much more blue than yellow or red light.

Rayleigh scattering seen over the ocean.

The process of scattering, or reradiating part of the electromagnetic wave by a charge oscillator, is fundamental to understanding the interaction of electromagnetic radiation with solids, liquids, or any matter that contains a very large number of charges and thus an enormous number of charge oscillators. This also explains why a substance that has charge oscillators of certain frequencies absorbs and emits radiation of those frequencies.

When electromagnetic radiation falls on a large collection of individual small charge oscillators, as in a piece of glass or metal or a brick wall, all of these oscillators perform oscillations in unison, following the beat of the electric wave. As a result, all the oscillators emit secondary radiation in unison (or coherently), and the total secondary radiation coming from the solid consists of the sum of all these secondary coherent electromagnetic waves. This sum total yields radiation that is reflected from the surface of the solid and radiation that goes into the solid at a certain angle with respect to the normal of (i.e., a line perpendicular to) the surface. The latter is the refracted radiation that may be attenuated (absorbed) on its way through the solid.

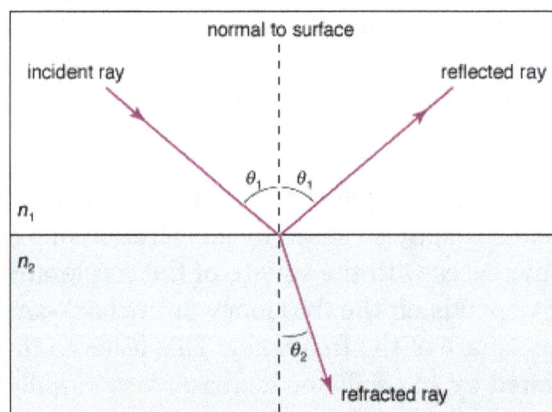

The law of refraction, or Snell's law, predicts the angle at which a light ray will bend, or refract, as it passes from one medium to another.

Superposition and Interference

When two electromagnetic waves of the same frequency superpose in space, the resultant electric and magnetic field strength of any point of space and time is the sum of the respective fields of the two waves. When one forms the sum, both the magnitude and the direction of the fields need be considered, which means that they sum like vectors. In the special case when two equally strong waves have their fields in the same direction in space and time (i.e., when they are in phase), the resultant field is twice that of each individual wave. The resultant intensity, being proportional to the square of the field strength, is therefore not two but four times the intensity of each of the two superposing waves.

By contrast, the superposition of a wave that has an electric field in one direction (positive) in space and time with a wave of the same frequency having an electric field in the opposite direction (negative) in space and time leads to cancellation and no resultant wave at all (zero intensity). Two waves of this sort are termed out of phase. The first example, that of in-phase superposition yielding four times the individual intensity, constitutes what is called constructive interference. The second example, that of out-of-phase superposition yielding zero intensity, is destructive interference. Since the resultant field at any point and time is the sum of all individual fields at that point and time, these arguments are easily extended to any number of superposing waves. One finds constructive, destructive, or partial interference for waves having the same frequency and given phase relationships.

Propagation and Coherence

Once generated, an electromagnetic wave is self-propagating because a time-varying electric field produces a time-varying magnetic field and vice versa. When an oscillating current in an antenna is switched on for, say, eight minutes, then the beginning of the electromagnetic train reaches the Sun just when the antenna is switched off because it takes a few seconds more than eight minutes for electromagnetic radiation to reach the Sun. This eight-minute wave train, which is as long as the Sun–Earth distance, then continues to travel with the speed of light past the Sun into the space beyond.

Except for radio waves transmitted by antennas that are switched on for many hours, most electromagnetic waves comes in many small pieces. The length and duration of a wave train are called coherence length and coherence time, respectively. Light from the Sun or from a lightbulb comes in many tiny bursts lasting about a millionth of a millionth of a second and having a coherence length of about one centimetre. The discrete radiant energy emitted by an atom as it changes its internal energy can have a coherence length several hundred times longer (one to 10 metres) unless the radiating atom is disturbed by a collision.

The time and space at which the electric and magnetic fields have a maximum value or are zero between the reversal of their directions are different for different wave trains. It is therefore clear that the phenomenon of interference can arise only from the superposition of part of a wave train with itself. This can be accomplished, for instance, with a half-transparent mirror that reflects half the intensity and transmits the other half of each of the billion billion wave trains of a given light source, say, a yellow sodium discharge lamp. One can allow one of these half beams to travel in direction A and the other in direction B, as shown in figure. By reflecting each half beam back, one can then superpose the two half beams and observe the resultant total. If one half beam has to travel a path $1/2$ wavelength or $3/2$ or $5/2$ wavelength longer than the other, then the superposition yields no light at all because the electric and magnetic fields of every half wave train in the two half beams point in opposite directions and their sum is therefore zero. The important point is that cancellation occurs between each half wave train and its mate. This is an example of destructive interference. By adjusting the path lengths A and B such that they are equal or differ by λ, 2λ, 3λ..., the electric and magnetic fields of each half wave train and its mate add when they are superposed. This is constructive interference, and, as a result, one sees strong light.

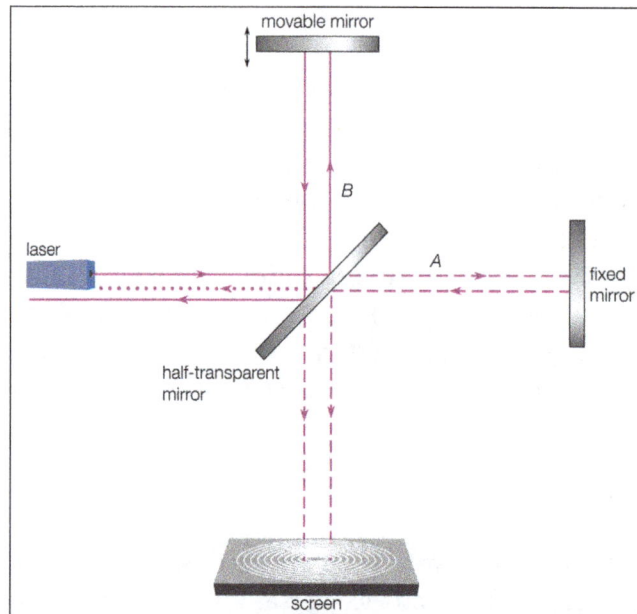

Michelson interferometer.

Speed of Electromagnetic Radiation and the Doppler Effect

Electromagnetic radiation—or, in modern terminology, the photons hv—always travels in free space with the universal speed c—i.e., the speed of light. This is actually a very puzzling situation which was first experimentally verified by Michelson and Edward Williams Morley, another American scientist, in 1887. It is the basic axiom of Albert Einstein's theory of relativity. Although there is no doubt that it is true, the situation is puzzling because it is so different from the behaviour of normal particles—that is to say, for little or not so little pieces of matter. When one chases behind a normal particle (e.g., an airplane) or moves from the opposite direction toward it, one certainly will measure very different speeds of the airplane relative to oneself. One would detect a very low relative speed in the first case and a very high one in the second. Moreover, a bullet shot forward from the airplane and another toward the back would appear to be moving with different speeds relative to oneself. This is not at all the case when one measures the speed of electromagnetic radiation: irrespective of one's motion or that of the source of the electromagnetic radiation, any measurement by a moving observer will result in the universal speed of light. This must be accepted as a fact of nature.

What happens to pitch or frequency when the source is moving toward the observers or away from them? It has been established from sound waves that the frequency is higher when a sound source is moving toward the observers and lower when it is moving away from them. This is the Doppler effect, named after the Austrian physicist Christian Doppler, who first described the phenomenon in 1842. Doppler predicted that the effect also occurs with electromagnetic radiation and suggested that it be used for measuring the relative speeds of stars. This explains why a characteristic blue light emitted, for example, by an excited helium atom as it changes from a higher to a lower internal energy state no longer appears blue when one looks at this light coming from helium atoms that move very rapidly away from Earth with, say, a galaxy. When the speed of such a galaxy away from Earth is high, the light may appear yellow; if the speed is still higher, it may appear red or even infrared. Hence, the speed of galaxies as well as of stars relative to Earth is measured from the Doppler shift of characteristic atomic radiation energies hv.

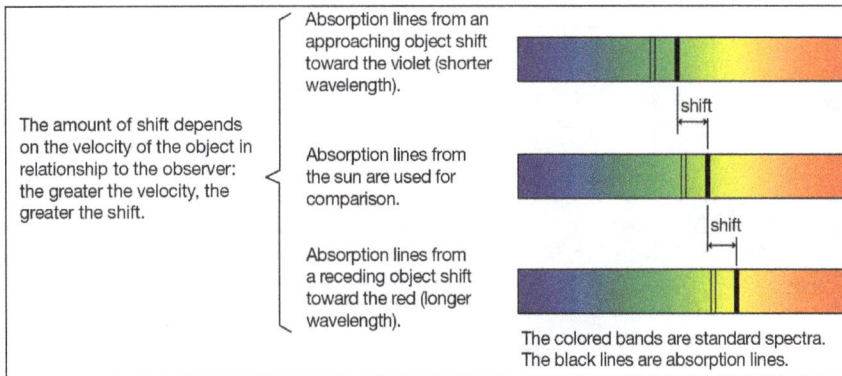

The amount of shift depends on the velocity of the object in relationship to the observer: the greater the velocity, the greater the shift.

Absorption lines from an approaching object shift toward the violet (shorter wavelength).

Absorption lines from the sun are used for comparison.

Absorption lines from a receding object shift toward the red (longer wavelength).

The colored bands are standard spectra. The black lines are absorption lines.

Doppler shift.

Forms of Electromagnetic Radiation

Electromagnetic radiation appears in a wide variety of forms and manifestations. Yet, these diverse phenomena are understood to comprise a single aspect of nature, following simple physical principles. Common to all forms is the fact that electromagnetic radiation interacts with and is generated by electric charges. The apparent differences in the phenomena arise from the question in which environment and under what circumstances can charges respond on the time scale of the frequency v of the radiation.

At smaller frequencies v (smaller than 10^{12} hertz), electric charges typically are the freely moving electrons in the metal components of antennas or the free electrons and ions in space that give rise to phenomena related to radio waves, radar waves, and microwaves. At higher frequencies (10^{12} to 5×10^{14} hertz), in the infrared region of the spectrum, the moving charges are primarily associated with the rotations and vibrations of molecules and the motions of atoms bonded together in materials. Electromagnetic radiation in the visible range to X-rays have frequencies that correspond to charges within atoms, whereas gamma rays are associated with frequencies of charges within atomic nuclei.

Radio Waves

Radio waves are used for wireless transmission of sound messages, or information, for communication, as well as for maritime and aircraft navigation. The information is imposed on the electromagnetic carrier wave as amplitude modulation (AM) or as frequency modulation (FM) or in digital form (pulse modulation). Transmission therefore involves not a single-frequency electromagnetic wave but rather a frequency band whose width is proportional to the information density. The width is about 10,000 Hz for telephone, 20,000 Hz for high-fidelity sound, and five megahertz (MHz = one million hertz) for high-definition television. This width and the decrease in efficiency of generating electromagnetic waves with decreasing frequency sets a lower frequency limit for radio waves near 10,000 Hz.

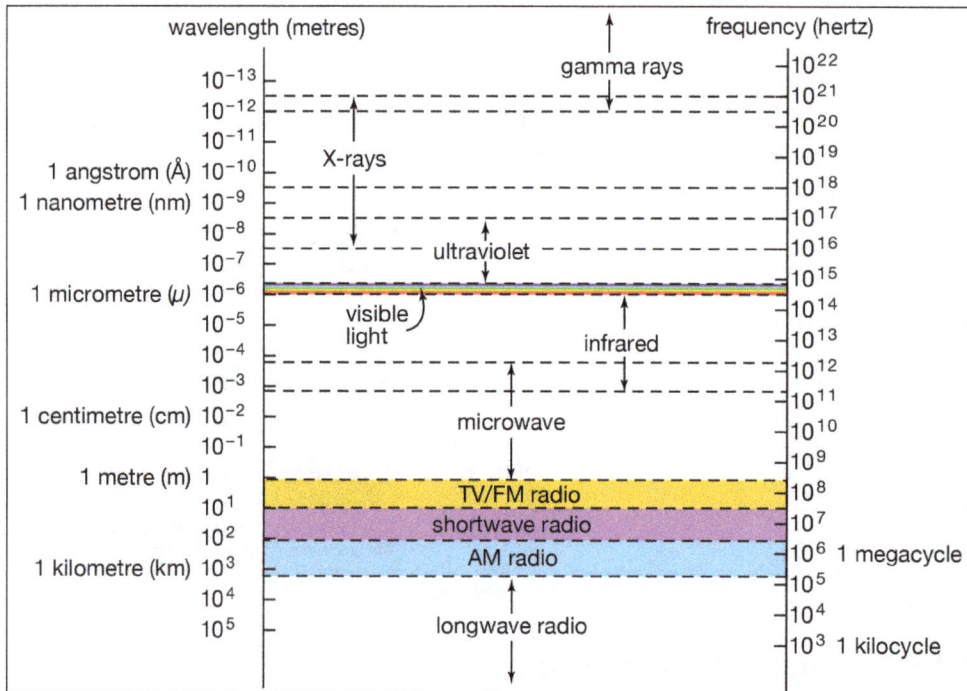

The electromagnetic spectrum.

Because electromagnetic radiation travels in free space in straight lines, late 19th-century scientists questioned the efforts of the Italian physicist and inventor Guglielmo Marconi to develop long-range radio. Earth's curvature limits the line-of-sight distance from the top of a 100-metre (330-foot) tower to about 30 km (19 miles). Marconi's unexpected success in transmitting messages over more than 2,000 km (1,200 miles) led to the discovery of the Kennelly-Heaviside layer, more commonly known as the ionosphere. This region is an approximately 300-km- (190-mile-) thick layer starting about 100 km (60 miles) above Earth's surface in which the atmosphere is partially ionized by ultraviolet light from the Sun, giving rise to enough electrons and ions to affect radio waves. Because of the Sun's involvement, the height, width, and degree of ionization of the stratified ionosphere vary from day to night and from summer to winter.

Radio waves transmitted by antennas in certain directions are bent or even reflected back to Earth by the ionosphere, as illustrated in Figure. They may bounce off Earth and be reflected by the ionosphere repeatedly, making radio transmission around the globe possible. Long-distance communication is further facilitated by the so-called ground wave. This form of electromagnetic wave closely follows Earth's surface, particularly over water, as a result of the wave's interaction with the terrestrial surface. The range of the ground wave (up to 1,600 km [1,000 miles]) and the bending and reflection of the sky wave by the ionosphere depend on the frequency of the waves. Under normal ionospheric conditions 40 MHz is the highest-frequency radio wave that can be reflected from the ionosphere. In order to accommodate the large band width of transmitted signals, television frequencies are necessarily higher than 40 MHz. Television transmitters must therefore be placed on high towers or on hilltops.

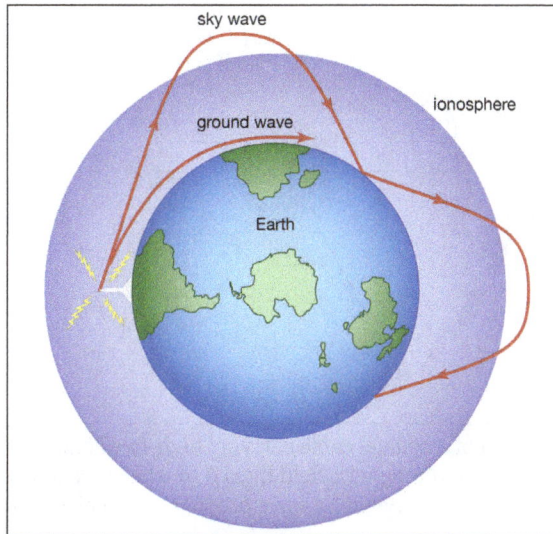

Radio-wave transmission reaching beyond line of sight by means of the sky wave reflected by the ionosphere and by means of the ground wave.

As a radio wave travels from the transmitting to the receiving antenna, it may be disturbed by reflections from buildings and other large obstacles. Disturbances arise when several such reflected parts of the wave reach the receiving antenna and interfere with the reception of the wave. Radio waves can penetrate nonconducting materials, such as wood, bricks, and concrete, fairly well. They cannot pass through electrical conductors, such as water or metals. Above $v = 40$ MHz, radio waves from deep space can penetrate Earth's atmosphere. This makes radio-astronomy observations with ground-based telescopes possible.

Whenever transmission of electromagnetic energy from one location to another is required with minimal energy loss and disturbance, the waves are confined to a limited region by means of wires, coaxial cables, and, in the microwave region, waveguides. Unguided or wireless transmission is naturally preferred when the locations of receivers are unspecified or too numerous, as in the case of radio and television communications. Cable television, as the name implies, is an exception. In this case electromagnetic radiation is transmitted by a coaxial cable system to users either from a community antenna or directly from broadcasting stations. The shielding of this guided transmission from disturbances provides high-quality signals.

Figure shows the electric field E (solid lines) and the magnetic field B (dashed lines) of an electromagnetic wave guided by a coaxial cable. There is a potential difference between the inner and outer conductors and so electric field lines E extend from one conductor to the other, represented here in cross section. The conductors carry opposite currents that produce the magnetic field lines B. The electric and magnetic fields are perpendicular to each other and perpendicular to the direction of propagation, as is characteristic of the electromagnetic waves illustrated in figure. At any cross section viewed, the directions of the E and B field lines change to their opposite with the frequency v of the radiation. This direction reversal of the fields does not change the direction of propagation along the conductors. The speed of propagation is again the universal speed of light if the region between the conductors consists of air or free space.

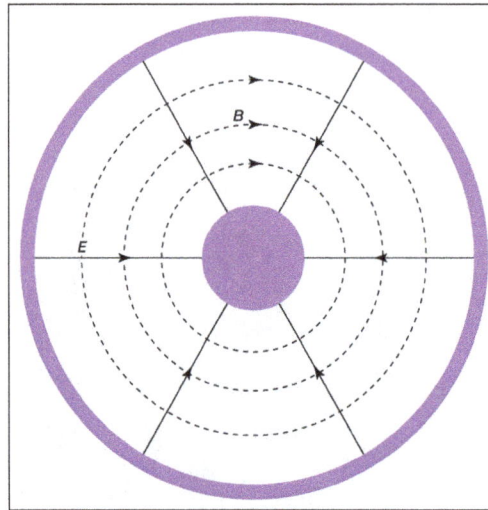

Cross section of a coaxial cable carrying high-frequency current.
Electric field lines E (solid) and magnetic field lines B (dashed) are mutually perpendicular
and perpendicular to the electromagnetic wave propagation, which is toward the viewer.

A combination of radio waves and strong magnetic fields is used by magnetic resonance imaging (MRI) to produce diagnostic pictures of parts of the human body and brain without apparent harmful effects. This imaging technique has thus found increasingly wider application in medicine.

Extremely low-frequency (ELF) waves are of interest for communications systems for submarines. The relatively weak absorption by seawater of electromagnetic radiation at low frequencies and the existence of prominent resonances of the natural cavity formed by Earth and the ionosphere make the range between 5 and 100 Hz attractive for this application.

Microwaves

The microwave region extends from 1,000 to 300,000 MHz (or 30 cm to 1 mm wavelength). Although microwaves were first produced and studied in 1886 by Hertz, their practical application had to await the invention of suitable generators, such as the klystron and magnetron.

Microwaves are the principal carriers of high-speed data transmissions between stations on Earth and also between ground-based stations and satellites and space probes. A system of synchronous satellites about 36,000 km above Earth is used for international broadband of all kinds of communications—e.g., television and telephone.

Microwave transmitters and receivers are parabolic dish antennas. They produce microwave beams whose spreading angle is proportional to the ratio of the wavelength of the constituent waves to the diameter of the dish. The beams can thus be directed like a searchlight. Radar beams consist of short pulses of microwaves. One can determine the distance of an airplane or ship by measuring the time it takes such a pulse to travel to the object and, after reflection, back to the radar dish antenna. Moreover, by making use of the change in frequency of the reflected wave pulse caused by the Doppler effect, one can measure the speed of objects. Microwave radar is therefore widely used for guiding airplanes and vessels and for detecting speeding motorists. Microwaves can penetrate clouds of smoke but are scattered by water droplets, so they are used for mapping meteorologic disturbances and in weather forecasting.

Microwaves play an increasingly wide role in heating and cooking food. They are absorbed by water and fat in foodstuffs (e.g., in the tissue of meats) and produce heat from the inside. In most cases, this reduces the cooking time a hundredfold. Such dry objects as glass and ceramics, on the other hand, are not heated in the process, and metal foils are not penetrated at all.

The heating effect of microwaves destroys living tissue when the temperature of the tissue exceeds 43 °C (109 °F). Accordingly, exposure to intense microwaves in excess of 20 milliwatts of power per square centimetre of body surface is harmful. The lens of the human eye is particularly affected by waves with a frequency of 3000 MHz, and repeated and extended exposure can result in cataracts. Radio waves and microwaves of far less power (microwatts per square centimetre) than the 10–20 milliwatts per square centimetre needed to produce heating in living tissue can have adverse effects on the electrochemical balance of the brain and the development of a fetus if these waves are modulated or pulsed at low frequencies between 5 and 100 hertz, which are of the same magnitude as brain wave frequencies.

Various types of microwave generators and amplifiers have been developed. Vacuum-tube devices, the klystron and the magnetron, continue to be used on a wide scale, especially for higher-power applications. Klystrons are primarily employed as amplifiers in radio relay systems and for dielectric heating, while magnetrons have been adopted for radar systems and microwave ovens. (For a detailed discussion of these devices, see electron tube.) Solid-state technology has yielded several devices capable of producing, amplifying, detecting, and controlling microwaves. Notable among these are the Gunn diode and the tunnel (or Esaki) diode. Another type of device, the maser (acronym for "microwave amplification by stimulated emission of radiation") has proved useful in such areas as radio astronomy, microwave radiometry, and long-distance communications.

Astronomers have discovered what appears to be natural masers in some interstellar clouds. Observations of radio radiation from interstellar hydrogen (H_2) and certain other molecules indicate amplification by the maser process. Also, microwave cosmic background radiation has been detected and is considered by many to be the remnant of the primeval fireball postulated by the big-bang cosmological model.

Infrared Radiation

Beyond the red end of the visible range but at frequencies higher than those of radar waves and microwaves is the infrared region of the electromagnetic spectrum, between frequencies of 10^{12} and 5×10^{14} Hz (or wavelengths from 0.1 to 7.5×10^{-5} cm). William Herschel, a German-born British musician and self-taught astronomer, discovered this form of radiation in 1800 by exploring, with the aid of a thermometer, sunlight dispersed into its colours by a glass prism. Infrared radiation is absorbed and emitted by the rotations and vibrations of chemically bonded atoms or groups of atoms and thus by many kinds of materials. For instance, window glass that is transparent to visible light absorbs infrared radiation by the vibration of its constituent atoms. Infrared radiation is strongly absorbed by water, as shown in figure, and by the atmosphere. Although invisible to the eye, infrared radiation can be detected as warmth by the skin. Nearly 50 percent of the Sun's radiant energy is emitted in the infrared region of the electromagnetic spectrum, with the rest primarily in the visible region.

Atmospheric haze and certain pollutants that scatter visible light are nearly transparent to parts of the infrared spectrum because the scattering efficiency increases with the fourth power of the frequency. Infrared photography of distant objects from the air takes advantage of this phenomenon. For the same reason, infrared astronomy enables researchers to observe cosmic objects through large clouds of interstellar dust that scatter infrared radiation substantially less than visible light. However, since water vapour, ozone, and carbon dioxide in the atmosphere absorb large parts of the infrared spectrum, many infrared astronomical observations are carried out at high altitude by balloons, rockets, aircraft, or spacecraft.

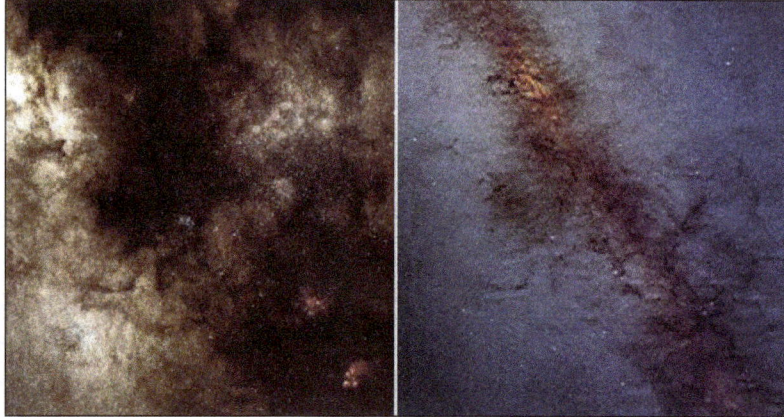

Central regions of the Milky Way Galaxy. The image on the left is in visible light, and the image on the right is in infrared; the marked difference between the two images shows how infrared radiation can penetrate galactic dust. The infrared image is part of the Two Micron All Sky Survey (2MASS), a survey of the entire sky in infrared light.

An infrared photograph of a landscape enhances objects according to their heat emission: blue sky and water appear nearly black, whereas green foliage and unexposed skin show up brightly. Infrared photography can reveal pathological tissue growths (thermography) and defects in electronic systems and circuits due to their increased emission of heat.

The infrared absorption and emission characteristics of molecules and materials yield important information about the size, shape, and chemical bonding of molecules and of atoms and ions in solids. The energies of rotation and vibration are quantized in all systems. The infrared radiation energy hv emitted or absorbed by a given molecule or substance is therefore a measure of the difference of some of the internal energy states. These in turn are determined by the atomic weight and molecular bonding forces. For this reason, infrared spectroscopy is a powerful tool for determining the internal structure of molecules and substances or, when such information is already known and tabulated, for identifying the amounts of those species in a given sample. Infrared spectroscopic techniques are often used to determine the composition and hence the origin and age of archaeological specimens and for detecting forgeries of art and other objects, which, when inspected under visible light, resemble the originals.

Infrared radiation plays an important role in heat transfer and is integral to the so-called greenhouse effect (see above The greenhouse effect of the atmosphere), influencing the thermal radiation budget of Earth on a global scale and affecting nearly all biospheric activity. Virtually every object at Earth's surface emits electromagnetic radiation primarily in the infrared region of the spectrum.

Artificial sources of infrared radiation include, besides hot objects, infrared light-emitting diodes (LEDs) and lasers. LEDs are small inexpensive optoelectronic devices made of such semiconducting materials as gallium arsenide. Infrared LEDs are employed as optoisolators and as light sources in some fibre-optics-based communications systems. Powerful optically pumped infrared lasers have been developed by using carbon dioxide and carbon monoxide. Carbon dioxide infrared lasers are used to induce and alter chemical reactions and in isotope separation. They also are employed in lidar systems. Other applications of infrared light include its use in the range finders of automatic self-focusing cameras, security alarm systems, and night-vision optical instruments.

Instruments for detecting infrared radiation include heat-sensitive devices such as thermocouple detectors, bolometers (some of these are cooled to temperatures close to absolute zero so that the thermal radiation of the detector system itself is greatly reduced), photovoltaic cells, and photoconductors. The latter are made of semiconductor materials (e.g., silicon and lead sulfide) whose electrical conductance increases when exposed to infrared radiation.

Visible Radiation

Visible light is the most familiar form of electromagnetic radiation and makes up that portion of the spectrum to which the eye is sensitive. This span is very narrow; the frequencies of violet light are only about twice those of red. The corresponding wavelengths extend from 7×10^{-5} cm (red) to 4×10^{-5} cm (violet). The energy of a photon from the centre of the visible spectrum (yellow) is $h\nu$ = 2.2 eV. This is one million times larger than the energy of a photon of a television wave and one billion times larger than that of radio waves in general.

Life on Earth could not exist without visible light, which represents the peak of the Sun's spectrum and close to one-half of all of its radiant energy. Visible light is essential for photosynthesis, which enables plants to produce the carbohydrates and proteins that are the food sources for animals. Coal and oil are sources of energy accumulated from sunlight in plants and microorganisms millions of years ago, and hydroelectric power is extracted from one step of the hydrologic cycle kept in motion by sunlight at the present time.

Considering the importance of visible sunlight for all aspects of terrestrial life, one cannot help being awed by the absorption spectrum of water in figure. The remarkable transparency of water centred in the narrow regime of visible light, indicated by vertical dashed lines in figure, is the result of the characteristic distribution of internal energy states of water. Absorption is strong toward the infrared on account of molecular vibrations and intermolecular oscillations. In the ultraviolet region, absorption of radiation is caused by electronic excitations. Light of frequencies having absorption coefficients larger than $\alpha = 10$ cm^{-1} cannot even reach the retina of the human eye, because its constituent liquid consists mainly of water that absorbs such frequencies of light.

Since the 1970s an increasing number of devices have been developed for converting sunlight into electricity. Unlike various conventional energy sources, solar energy does not become depleted by use and does not pollute the environment. Two branches of development may be noted—namely, photothermal and photovoltaic technologies. In photothermal devices, sunlight is used to heat a substance, as, for example, water, to produce steam with which to drive a generator. Photovoltaic devices, on the other hand, convert the energy in sunlight directly to electricity by use of the photovoltaic effect in a semiconductor junction. Solar panels consisting of photovoltaic devices made of gallium arsenide have conversion efficiencies of more than 20 percent and are used to provide electric power in many satellites and space probes. Solar cells have replaced dry-cell batteries in some portable electronic instruments, and solar energy power stations of more than 500 mega-watts capacity have been built.

The intensity and spectral composition of visible light can be measured and recorded by essentially any process or property that is affected by light. Detectors make use of a photographic process based on silver halide, the photoemission of electrons from metal surfaces, the generation of electric current in a photovoltaic cell, and the increase in electrical conduction in semiconductors.

Glass fibres constitute an effective means of guiding and transmitting light. A beam of light is confined by total internal reflection to travel inside such an optical fibre, whose thickness may be anywhere between one hundredth of a millimetre and a few millimetres. Many thin optical fibres can be combined into bundles to achieve image reproduction. The flexibility of these fibres or fibre bundles permits their use in medicine for optical exploration of internal organs. Optical fibres connecting the continents provide the capability to transmit substantially larger amounts of information than other systems of international telecommunications. Another advantage of optical fibre communication systems is that transmissions cannot easily be intercepted and are not disturbed by lower atmospheric and stratospheric disturbances.

Optical fibres integrated with miniature semiconductor lasers and light-emitting diodes, as well as with light detector arrays and photoelectronic imaging and recording materials, form the building blocks of a new optoelectronics industry. Some familiar commercial products are optoelectronic copying machines, laser printers, compact disc players, optical recording media, and optical disc mass-storage systems of exceedingly high bit density.

Ultraviolet Radiation

The German physicist Johann Wilhelm Ritter, having learned of Herschel's discovery of infrared waves, looked beyond the violet end of the visible spectrum of the Sun and found (in 1801) that there exist invisible rays that darken silver chloride even more efficiently than visible light. This spectral region extending between visible light and X-rays is designated ultraviolet. Sources of this form of electromagnetic radiation are hot objects like the Sun, synchrotron radiation sources, mercury or xenon arc lamps, and gaseous discharge tubes filled with gas atoms (e.g., mercury, deuterium, or hydrogen) that have internal electron energy levels which correspond to the photons of ultraviolet light.

When ultraviolet light strikes certain materials, it causes them to fluoresce—i.e., they emit electromagnetic radiation of lower energy, such as visible light. The spectrum of fluorescent light is characteristic of a material's composition and thus can be used for screening minerals, detecting bacteria in spoiled food, identifying pigments, or detecting forgeries of artworks and other objects (the aged surfaces of ancient marble sculptures, for instance, fluoresce yellow-green, whereas a freshly cut marble surface fluoresces bright violet).

Optical instruments for the ultraviolet region are made of special materials, such as quartz, certain silicates, and metal fluorides, which are transparent at least in the near ultraviolet. Far-ultraviolet radiation is absorbed by nearly all gases and materials and thus requires reflection optics in vacuum chambers.

Ultraviolet radiation is detected by photographic plates and by means of the photoelectric effect in photomultiplier tubes. Also, ultraviolet radiation can be converted to visible light by fluorescence before detection.

The relatively high energy of ultraviolet light gives rise to certain photochemical reactions. This characteristic is exploited to produce cyanotype impressions on fabrics and for blueprinting design drawings. Here, the fabric or paper is treated with a mixture of chemicals that react upon exposure to ultraviolet light to form an insoluble blue compound. Electronic excitations caused by ultraviolet radiation also produce changes in the colour and transparency of photosensitive and photochromic glasses. Photochemical and photostructural changes in certain polymers constitute the basis for photolithography and the processing of the microelectronic circuits.

Although invisible to the eyes of humans and most vertebrates, near-ultraviolet light can be seen by many insects. Butterflies and many flowers that appear to have identical colour patterns under visible light are distinctly different when viewed under the ultraviolet rays perceptible to insects.

An evening primrose (Oenothera biennis) seen (top) in visible light and (bottom) in ultraviolet light; the latter reveals nectar-guide patterns that are discernible to the moth pollinating this flower but not to the human eye.

An important difference between ultraviolet light and electromagnetic radiation of lower frequencies is the ability of the former to ionize, meaning that it can knock an electron out from atoms and molecules. All high-frequency electromagnetic radiation beyond the visible—i.e., ultraviolet light, X-rays, and gamma rays—is ionizing and therefore harmful to body tissues, living cells, and DNA (deoxyribonucleic acid). The harmful effects of ultraviolet light to humans and larger animals are mitigated by the fact that this form of radiation does not penetrate much further than the skin.

The body of a sunbather is struck by 10^{21} photons every second, and 1 percent of these, or more than a billion billion per second, are photons of ultraviolet radiation. Tanning and natural body pigments help to protect the skin to some degree, preventing the destruction of skin cells by ultraviolet light. Nevertheless, overexposure to the ultraviolet component of sunlight can cause skin cancer, cataracts of the eyes, and damage to the body's immune system. Fortunately, a layer of ozone (O_3) in the stratosphere absorbs the most-damaging ultraviolet rays, which have wavelengths of 2000 and 2900 angstroms (one angstrom [Å] = 10^{-10} metre), and attenuates those with wavelengths between 2900 and 3150 Å. Without this protective layer of ozone, life on Earth would not be possible. The ozone layer is produced at an altitude of about 10 to 50 km (6 to 30 miles) above Earth's surface by a reaction between upward-diffusing molecular oxygen (O_2) and downward-diffusing ionized atomic oxygen (O^+). In the late 20th century this life-protecting stratospheric ozone layer was reduced by chlorine atoms in chlorofluorocarbon (or Freon) gases released into the atmosphere by aerosol propellants, air-conditioner coolants, solvents used in the manufacture of electronic components, and other sources. Limits were placed on the sale of ozone-depleting chemicals, and the ozone layer was expected to recover eventually.

Ionized atomic oxygen, nitrogen, and nitric oxide are produced in the upper atmosphere by absorption of solar ultraviolet radiation. This ionized region is the ionosphere, which affects radio communications and reflects and absorbs radio waves of frequencies below 40 MHz.

X-rays

The German physicist Wilhelm Conrad Röntgen discovered X-rays in 1895 by accident while studying cathode rays in a low-pressure gas discharge tube. (A few years later J.J. Thomson of England showed that cathode rays were electrons emitted from the negative electrode [cathode] of the discharge tube.) Röntgen noticed the fluorescence of a barium platinocyanide screen that happened to lie near the discharge tube. He traced the source of the hitherto undetected form of radiation to the point where the cathode rays hit the wall of the discharge tube, and he mistakenly concluded from his inability to observe reflection or refraction that his new rays were unrelated to light. Because of his uncertainty about their nature, he called them X-radiation. This early failure can be attributed to the very short wavelengths of X-rays (10^{-8} to 10^{-11} cm), which correspond to photon energies from 200 to 100,000 eV. In 1912 another German physicist, Max von Laue, realized that the regular arrangement of atoms in crystals should provide a natural grating of the right spacing (about 10^{-8} cm) to produce an interference pattern on a photographic plate when X-rays pass through such a crystal. The success of this experiment, carried out by Walter Friedrich and Paul Knipping, not only identified X-rays with electromagnetic radiation but also initiated the use of X-rays for studying the detailed atomic structure of crystals. The interference of X-rays diffracted in certain directions from crystals in so-called X-ray diffractometers, in turn, permits the dissection of X-rays into their different frequencies, just as a prism diffracts and spreads the various colours of light. The spectral composition and characteristic frequencies of X-rays emitted by a given X-ray source can thus be measured. As in optical spectroscopy, the X-ray photons emitted correspond to the differences of the internal electronic energies in atoms and molecules. Because of their much higher energies, however, X-ray photons are associated with the inner-shell electrons close to the atomic nuclei, whereas optical absorption and emission are related to the outermost electrons in atoms or in materials in general. Since the outer electrons are used for chemical bonding while the energies of inner-shell electrons remain essentially unaffected by atomic bonding, the identity and quantity of elements that make up a material are more accurately determined by the emission, absorption, or fluorescence of X-rays than of photons of visible or ultraviolet light.

The contrast between body parts in medical X-ray photographs (radiographs) is produced by the different scattering and absorption of X-rays by bones and tissues. Within months of Röntgen's discovery of X-rays and his first X-ray photograph of his wife's hand, this form of electromagnetic radiation became indispensable in orthopedic and dental medicine. The use of X-rays for obtaining images of the body's interior has undergone considerable development over the years and has culminated in the highly sophisticated procedure known as computed tomography.

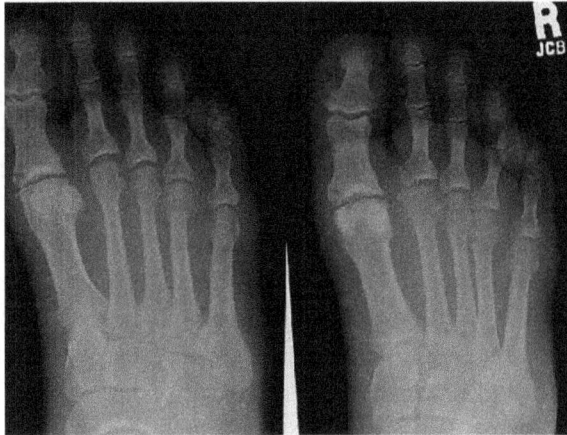

Metatarsal: X-ray showing metatarsal bones of a foot.

Not with standing their usefulness in medical diagnosis, the ability of X-rays to ionize atoms and molecules and their penetrating power make them a potential health hazard. Exposure of body cells and tissue to large doses of such ionizing radiation can result in abnormalities in DNA that may lead to cancer and birth defects.

X-rays are produced in X-ray tubes by the deceleration of energetic electrons (bremsstrahlung) as they hit a metal target or by accelerating electrons moving at relativistic velocities in circular orbits (synchrotron radiation; see above Continuous spectra of electromagnetic radiation). They are detected by their photochemical action in photographic emulsions or by their ability to ionize gas atoms. Every X-ray photon produces a burst of electrons and ions, resulting in a current pulse. By counting the rate of such current pulses per second, the intensity of a flux of X-rays can be measured. Instruments used for this purpose are called Geiger counters.

X-ray astronomy has revealed very strong sources of X-rays in deep space. In the Milky Way Galaxy, of which the solar system is a part, the most-intense sources are certain double-star systems in which one of the two stars is thought to be either a compact neutron star or a black hole. The ionized gas of the circling companion star falls by gravitation into the compact star, generating X-rays that may be more than 1,000 times as intense as the total amount of light emitted by the Sun. At the moment of their explosion, supernovae emit a good fraction of their energy in a burst of X-rays.

Gamma Rays

Six years after the discovery of radioactivity by Henri Becquerel of France, the New Zealand-born British physicist Ernest Rutherford found that three different kinds of radiation are emitted in the decay of radioactive substances; these he called alpha, beta, and gamma rays in sequence of their ability to penetrate matter. The alpha particles were found to be identical with the nuclei of helium atoms, and the beta rays were identified as electrons. In 1912 it was shown that the much more penetrating gamma rays have all the properties of very energetic electromagnetic radiation, or photons. Gamma-ray photons are between 10,000 and 10,000,000 times more energetic than the photons of visible light when they originate from radioactive atomic nuclei. Gamma rays with a million million times higher energy make up a very small part of the cosmic rays that reach Earth from supernovae or from other galaxies. The origin of the most-energetic gamma rays is not yet known.

During radioactive decay, an unstable nucleus usually emits alpha particles, electrons, gamma rays, and neutrinos spontaneously. In nuclear fission, the unstable nucleus breaks into fragments, which are themselves complex nuclei, along with such particles as neutrons and protons. The resultant stable nuclei or nuclear fragments are usually in a highly excited state and then reach their low-energy ground state by emitting one or more gamma rays. Such a decay scheme is shown schematically in figure for the unstable nucleus sodium-24 (^{24}Na). Much of what is known about the internal structure and energies of nuclei has been obtained from the emission or resonant absorption of gamma rays by nuclei. Absorption of gamma rays by nuclei can cause them to eject neutrons or alpha particles or it can even split a nucleus like a bursting bubble in what is called photodisintegration. A gamma particle hitting a hydrogen nucleus (that is, a proton), for example, produces a positive pi-meson and a neutron or a neutral pi-meson and a proton. Neutral pi-mesons, in turn, have a very brief mean life of 1.8×10^{-16} second and decay into two gamma rays of energy $h\nu \approx 70$ MeV. When an energetic gamma ray $h\nu > 1.02$ MeV passes a nucleus, it may disappear while creating an electron–positron pair. Gamma photons interact with matter by discrete elementary processes that include resonant absorption, photodisintegration, ionization, scattering (Compton scattering), or pair production.

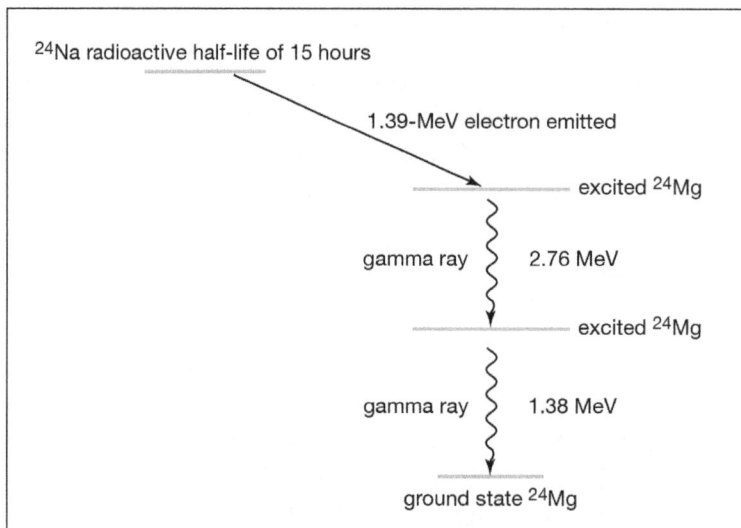

^{24}Na radioactive half-life of 15 hours

1.39-MeV electron emitted

excited ^{24}Mg

gamma ray   2.76 MeV

excited ^{24}Mg

gamma ray   1.38 MeV

ground state ^{24}Mg

Decay scheme of a radioactive sodium-24 (^{24}Na) nucleus. With a half-life of 15 hours, it decays by beta decay to an excited magnesium-24 (^{24}Mg) nucleus. Two gamma rays are rapidly emitted and the excitation energy is carried off, whereby the stable ground state of magnesium-24 is reached.

Gamma rays are detected by their ability to ionize gas atoms or to create electron–hole pairs in semiconductors or insulators. By counting the rate of charge pulses or voltage pulses or by measuring the scintillation of the light emitted by the subsequently recombining electron–hole pairs, one can determine the number and energy of gamma rays striking an ionization detector or scintillation counter.

Both the specific energy of the gamma-ray photon emitted as well as the half-life of the specific radioactive decay process that yields the photon identify the type of nuclei at hand and their concentrations. By bombarding stable nuclei with neutrons, one can artificially convert more than 70 different stable nuclei into radioactive nuclei and use their characteristic gamma emission for purposes of identification, for impurity analysis of metallurgical specimens (neutron-activation analysis), or as radioactive tracers with which to determine the functions or malfunctions of human organs, to follow the life cycles of organisms, or to determine the effects of chemicals on biological systems and plants.

The great penetrating power of gamma rays stems from the fact that they have no electric charge and thus do not interact with matter as strongly as do charged particles. Because of their penetrating power gamma rays can be used for radiographing holes and defects in metal castings and other structural parts. At the same time, this property makes gamma rays extremely hazardous. The lethal effect of this form of ionizing radiation makes it useful for sterilizing medical supplies that cannot be sanitized by boiling or for killing organisms that cause food spoilage. More than 50 percent of the ionizing radiation to which humans are exposed comes from natural radon gas, which is an end product of the radioactive decay chain of natural radioactive substances in minerals. Radon escapes from the ground and enters the environment in varying amounts.

References

- Srednicki, Mark A. (2007). Quantum field theory. Cambridge, [England] ; New York [NY.]: Cambridge University Press. ISBN 978-0-521-86449-7

- Electrodynamics, electrical, engineering, science, encyclopedia: infoplease.com, Retrieved 14 July, 2019

- Serway, Raymond A.; Jewett, John W., Jr. (2004). Physics for scientists and engineers, with modern physics. Belmont, [CA.]: Thomson Brooks/Cole. ISBN 0-534-40846-X

- Joseph Henry". Distinguished Members Gallery, National Academy of Sciences. Archived from the original on 2013-12-13. Retrieved 2006-11-30

- Raymond A. Serway, John W. Jewett (2006). Principles of Physics. Thomson Brooks/Cole. P. 807. ISBN 978-0-534-49143-7

- Liu, Changli (2017). "Explanation on Overdetermination of Maxwell's Equations". Physics and Engineering. 27 (3): 7–9. Arxiv:1002.0892. Bibcode:2010arxiv1002.0892L. Doi:10.3969/j.issn.1009-7104.2017.03.002

- Spencer, James N.; et al. (2010). Chemistry: Structure and Dynamics. John Wiley & Sons. P. 78. ISBN 97804 70587119

- electromagnetic-radiation,science: britannica.com, Retrieved 11 January, 2019

5

Electrical Circuits

The paths used for transmitting electric current are known as electrical circuits. Some of the theorems and laws studied in relation to electrical circuits are Ohm's law, Kirchhoff's circuit laws, Thévenin's theorem, Norton's theorem and superposition theorem. This chapter closely examines these major concepts associated with electric circuits.

Electric circuit is the path for transmitting electric current. An electric circuit includes a device that gives energy to the charged particles constituting the current, such as a battery or a generator; devices that use current, such as lamps, electric motors, or computers; and the connecting wires or transmission lines. Two of the basic laws that mathematically describe the performance of electric circuits are Ohm's law and Kirchhoff's rules.

Electric circuits are classified in several ways. A direct-current circuit carries current that flows only in one direction. An alternating-current circuit carries current that pulsates back and forth many times each second, as in most household circuits. A series circuit comprises a path along which the whole current flows through each component. A parallel circuit comprises branches so that the current divides and only part of it flows through any branch. The voltage, or potential difference, across each branch of a parallel circuit is the same, but the currents may vary. In a home electrical circuit, for instance, the same voltage is applied across each light or appliance, but each of these loads draws a different amount of current, according to its power requirements. A number of similar batteries connected in parallel provides greater current than a single battery, but the voltage is the same as for a single battery.

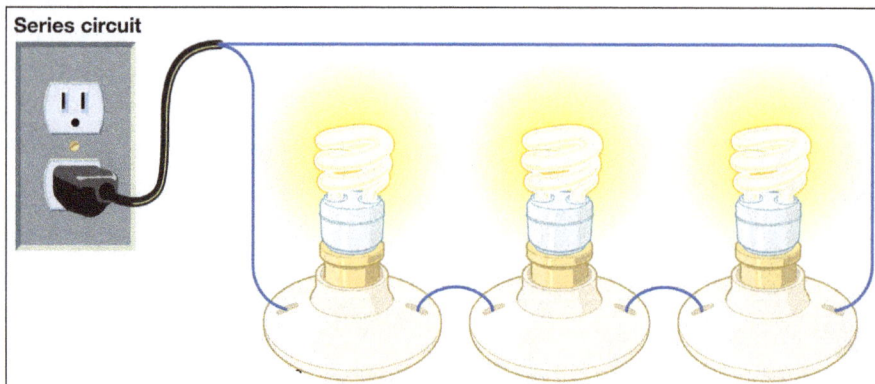

A series circuit.

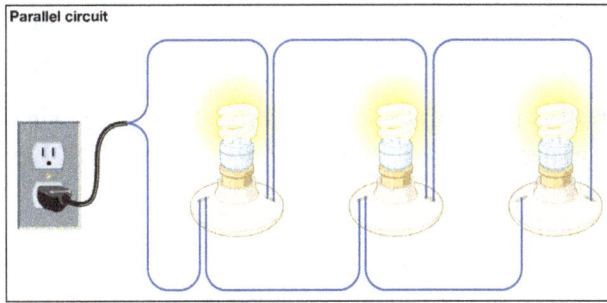

A parallel circuit.

The network of transistors, transformers, capacitors, connecting wires, and other electronic components within a single device such as a radio is also an electric circuit. Such complex circuits may be made up of one or more branches in combinations of series and series-parallel arrangements.

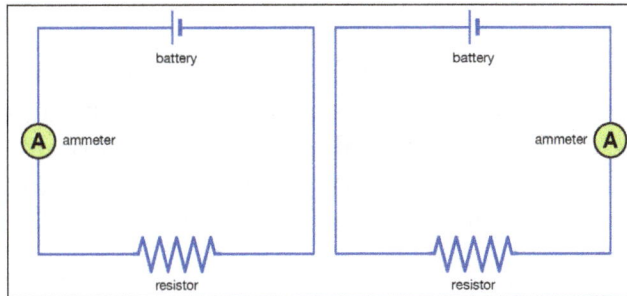

Ammeter: Two diagrams showing an ammeter connected to a simple circuit in two different positions.

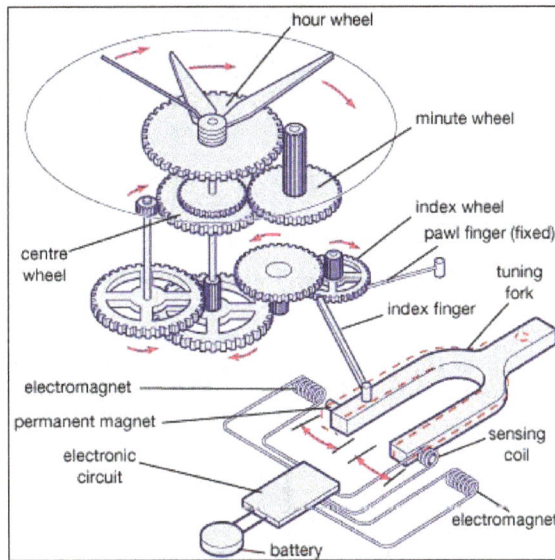

Typical components of an electrically powered watch.

Electric Current

An electric current is the rate of flow of electric charge past a point or region. An electric current is said to exist when there is a net flow of electric charge through a region. In electric circuits this charge is often carried by electrons moving through a wire. It can also be carried by ions in an electrolyte, or by both ions and electrons such as in an ionized gas (plasma).

The SI unit of electric current is the ampere, which is the flow of electric charge across a surface at the rate of one coulomb per second. The ampere (symbol: A) is an SI base unit Electric current is measured using a device called an ammeter.

Electric currents cause Joule heating, which creates light in incandescent light bulbs. They also create magnetic fields, which are used in motors, inductors and generators.

The moving charged particles in an electric current are called charge carriers. In metals, one or more electrons from each atom are loosely bound to the atom, and can move freely about within the metal. These conduction electrons are the charge carriers in metal conductors.

Symbol

The conventional symbol for current is I, which originates from the French phrase *intensité du courant*, (current intensity). Current intensity is often referred to simply as *current*. The I symbol was used by André-Marie Ampère, after whom the unit of electric current is named, in formulating Ampère's force law. The notation travelled from France to Great Britain, where it became standard, although at least one journal did not change from using C to I until 1896.

Conventions

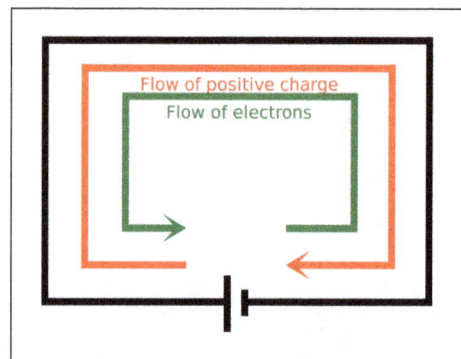

The electrons, the charge carriers in an electrical circuit, flow in the opposite direction of the conventional electric current.

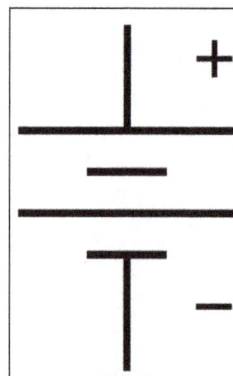

The symbol for a battery in a circuit diagram.

In a conductive material, the moving charged particles that constitute the electric current are called charge carriers. In metals, which make up the wires and other conductors in most electrical circuits, the positively charged atomic nuclei of the atoms are held in a fixed position, and the negatively charged electrons are the charge carriers, free to move about in the metal. In other materials, notably the semiconductors, the charge carriers can be positive *or* negative, depending on the dopant used. Positive and negative charge carriers may even be present at the same time, as happens in an electrolyte in an electrochemical cell.

A flow of positive charges gives the same electric current, and has the same effect in a circuit, as an equal flow of negative charges in the opposite direction. Since current can be the flow of either positive or negative charges, or both, a convention is needed for the direction of current that is independent of the type of charge carriers. The direction of *conventional current* is arbitrarily defined as the same direction as positive charges flow.

Since electrons, the charge carriers in metal wires and most other parts of electric circuits, have a negative charge, as a consequence, they flow in the opposite direction of conventional current flow in an electrical circuit.

Reference Direction

Since the current in a wire or component can flow in either direction, when a variable I is defined to represent that current, the direction representing positive current must be specified, usually by an arrow on the circuit schematic diagram. This is called the *reference direction* of current I. If the current flows in the opposite direction, the variable I has a negative value.

When analyzing electrical circuits, the actual direction of current through a specific circuit element is usually unknown. Consequently, the reference directions of currents are often assigned arbitrarily. When the circuit is solved, a negative value for the variable means that the actual direction of current through that circuit element is opposite that of the chosen reference direction. In electronic circuits, the reference current directions are often chosen so that all currents are toward ground. This often corresponds to the actual current direction, because in many circuits the power supply voltage is positive with respect to ground.

Occurrences

Natural observable examples of electrical current include lightning, static electric discharge, and the solar wind, the source of the polar auroras.

Man-made occurrences of electric current include the flow of conduction electrons in metal wires such as the overhead power lines that deliver electrical energy across long distances and the smaller wires within electrical and electronic equipment. Eddy currents are electric currents that occur in conductors exposed to changing magnetic fields. Similarly, electric currents occur, particularly in the surface, of conductors exposed to electromagnetic waves. When oscillating electric currents flow at the correct voltages within radio antennas, radio waves are generated.

In electronics, other forms of electric current include the flow of electrons through resistors or through the vacuum in a vacuum tube, the flow of ions inside a battery or a neuron, and the flow of holes within metals and semiconductors.

Current Measurement

Current can be measured using an ammeter.

Electric current can be directly measured with a galvanometer, but this method involves breaking the electrical circuit, which is sometimes inconvenient.

Current can also be measured without breaking the circuit by detecting the magnetic field associated with the current. Devices, at the circuit level, use various techniques to measure current:

- Shunt resistors.

- Hall effect current sensor transducers.

- Transformers (however DC cannot be measured).

- Magnetoresistive field sensors.

- Rogowski coils.

- current clamps.

Resistive Heating

Joule heating, also known as *ohmic heating* and *resistive heating*, is the process of power dissipation by which the passage of an electric current through a conductor increases the internal energy of the conductor, converting thermodynamic work into heat. The phenomenon was first studied by James Prescott Joule in 1841. Joule immersed a length of wire in a fixed mass of water and measured the temperature rise due to a known current through the wire for a 30 minute period. By varying the current and the length of the wire he deduced that the heat produced was proportional to the square of the current multiplied by the electrical resistance of the wire.

$$P \propto I^2 R$$

This relationship is known as Joule's Law. The SI unit of energy was subsequently named the joule and given the symbol J. The commonly known SI unit of power, the watt (symbol: W), is equivalent to one joule per second.

Conduction Mechanisms in Various Media

In metallic solids, electric charge flows by means of electrons, from lower to higher electrical potential. In other media, any stream of charged objects (ions, for example) may constitute an electric current. To provide a definition of current independent of the type of charge carriers, *conventional current* is defined as moving in the same direction as the positive charge flow. So, in metals where the charge carriers (electrons) are negative, conventional current is in the opposite direction as the electrons. In conductors where the charge carriers are positive, conventional current is in the same direction as the charge carriers.

In a vacuum, a beam of ions or electrons may be formed. In other conductive materials, the electric current is due to the flow of both positively and negatively charged particles at the same time. In still others, the current is entirely due to positive charge flow. For example, the electric currents in electrolytes are flows of positively and negatively charged ions. In a common lead-acid electro-chemical cell, electric currents are composed of positive hydronium ions flowing in one direction, and negative sulfate ions flowing in the other. Electric currents in sparks or plasma are flows of electrons as well as positive and negative ions. In ice and in certain solid electrolytes, the electric current is entirely composed of flowing ions.

Metals

In a metal, some of the outer electrons in each atom are not bound to the individual atom as they are in insulating materials, but are free to move within the metal lattice. These conduction electrons can serve as charge carriers, carrying a current. Metals are particularly conductive because there are many of these free electrons, typically one per atom in the lattice. With no external electric field applied, these electrons move about randomly due to thermal energy but, on average, there is zero net current within the metal. At room temperature, the average speed of these random motions is 10^6 metres per second. Given a surface through which a metal wire passes, electrons move in both directions across the surface at an equal rate.

When a metal wire is connected across the two terminals of a DC voltage source such as a battery, the source places an electric field across the conductor. The moment contact is made, the free electrons of the conductor are forced to drift toward the positive terminal under the influence of this field. The free electrons are therefore the charge carrier in a typical solid conductor.

For a steady flow of charge through a surface, the current I (in amperes) can be calculated with the following equation:

$$I = \frac{Q}{t},$$

where Q is the electric charge transferred through the surface over a time t. If Q and t are measured in coulombs and seconds respectively, I is in amperes.

More generally, electric current can be represented as the rate at which charge flows through a given surface as:

$$I = \frac{dQ}{dt}.$$

Electrolytes

Electric currents in electrolytes are flows of electrically charged particles (ions). For example, if an electric field is placed across a solution of Na^+ and Cl^- (and conditions are right) the sodium ions move towards the negative electrode (cathode), while the chloride ions move towards the positive electrode (anode). Reactions take place at both electrode surfaces, neutralizing each ion.

Water-ice and certain solid electrolytes called proton conductors contain positive hydrogen ions ('protons') that are mobile. In these materials, electric currents are composed of moving 'protons', as opposed to the moving electrons in metals.

In certain electrolyte mixtures, brightly coloured ions are the moving electric charges. The slow progress of the colour makes the current visible.

Gases and Plasmas

In air and other ordinary gases below the breakdown field, the dominant source of electrical conduction is via relatively few mobile ions produced by radioactive gases, ultraviolet light, or cosmic rays. Since the electrical conductivity is low, gases are dielectrics or insulators. However, once the applied electric field approaches the breakdown value, free electrons become sufficiently accelerated by the electric field to create additional free electrons by colliding, and ionizing, neutral gas atoms or molecules in a process called avalanche breakdown. The breakdown process forms a plasma that contains enough mobile electrons and positive ions to make it an electrical conductor. In the process, it forms a light emitting conductive path, such as a spark, arc or lightning.

Plasma is the state of matter where some of the electrons in a gas are stripped or "ionized" from their molecules or atoms. A plasma can be formed by high temperature, or by application of a high electric or alternating magnetic field as noted above. Due to their lower mass, the electrons in a plasma accelerate more quickly in response to an electric field than the heavier positive ions, and hence carry the bulk of the current. The free ions recombine to create new chemical compounds (for example, breaking atmospheric oxygen into single oxygen [$O_2 \rightarrow 2O$], which then recombine creating ozone [O_3]).

Vacuum

Since a "perfect vacuum" contains no charged particles, it normally behaves as a perfect insulator. However, metal electrode surfaces can cause a region of the vacuum to become conductive by injecting free electrons or ions through either field electron emission or thermionic emission. Thermionic emission occurs when the thermal energy exceeds the metal's work function, while field electron emission occurs when the electric field at the surface of the metal is high enough to cause tunneling, which results in the ejection of free electrons from the metal into the vacuum. Externally heated electrodes are often used to generate an electron cloud as in the filament or indirectly heated cathode of vacuum tubes. Cold electrodes can also spontaneously produce electron clouds via thermionic emission when small incandescent regions (called cathode spots or anode spots) are formed. These are incandescent regions of the electrode surface that are created by a localized high current. These regions may be initiated by field electron emission, but are then sustained by localized thermionic emission once a vacuum arc forms. These small electron-emitting regions can form quite rapidly, even explosively, on a metal surface subjected to a high electrical field. Vacuum tubes and sprytrons are some of the electronic switching and amplifying devices based on vacuum conductivity.

Semiconductor

In a semiconductor it is sometimes useful to think of the current as due to the flow of positive "holes" (the mobile positive charge carriers that are places where the semiconductor crystal is missing a valence electron). This is the case in a p-type semiconductor. A semiconductor has electrical conductivity intermediate in magnitude between that of a conductor and an insulator. This means a conductivity roughly in the range of 10^{-2} to 10^4 siemens per centimeter (S·cm^{-1}).

In the classic crystalline semiconductors, electrons can have energies only within certain bands (i.e. ranges of levels of energy). Energetically, these bands are located between the energy of the ground state, the state in which electrons are tightly bound to the atomic nuclei of the material, and the free electron energy, the latter describing the energy required for an electron to escape entirely from the material. The energy bands each correspond to many discrete quantum states of the electrons, and most of the states with low energy (closer to the nucleus) are occupied, up to a particular band called the *valence band*. Semiconductors and insulators are distinguished from metals because the valence band in any given metal is nearly filled with electrons under usual operating conditions, while very few (semiconductor) or virtually none (insulator) of them are available in the *conduction band*, the band immediately above the valence band.

The ease of exciting electrons in the semiconductor from the valence band to the conduction band depends on the band gap between the bands. The size of this energy band gap serves as an arbitrary dividing line (roughly 4 eV) between semiconductors and insulators.

With covalent bonds, an electron moves by hopping to a neighboring bond. The Pauli exclusion principle requires that the electron be lifted into the higher anti-bonding state of that bond. For delocalized states, for example in one dimension – that is in a nanowire, for every energy there is a state with electrons flowing in one direction and another state with the electrons flowing in the other. For a net current to flow, more states for one direction than for the other direction must be occupied. For this to occur, energy is required, as in the semiconductor the next higher states lie above the band gap. Often this is stated as: full bands do not contribute to the electrical conductivity. However, as a semiconductor's temperature rises above absolute zero, there is more energy in the semiconductor to spend on lattice vibration and on exciting electrons into the conduction band. The current-carrying electrons in the conduction band are known as *free electrons*, though they are often simply called *electrons* if that is clear in context.

Current Density and Ohm's Law

Current density is the rate at which charge passes through a chosen unit area. It is defined as a vector whose magnitude is the current per unit cross-sectional area. As discussed in Reference direction, the direction is arbitrary. Conventionally, if the moving charges are positive, then the current density has the same sign as the velocity of the charges. For negative charges, the sign of the current density is opposite to the velocity of the charges. In SI units, current density (symbol: j) is expressed in the SI base units of amperes per square metre.

In linear materials such as metals, and under low frequencies, the current density across the conductor surface is uniform. In such conditions, Ohm's law states that the current is directly proportional to the potential difference between two ends (across) of that metal (ideal) resistor (or other ohmic device):

$$I = \frac{V}{R},$$

where I is the current, measured in amperes; V is the potential difference, measured in volts; and R is the resistance, measured in ohms. For alternating currents, especially at higher frequencies, skin effect causes the current to spread unevenly across the conductor cross-section, with higher density near the surface, thus increasing the apparent resistance.

Drift Speed

The mobile charged particles within a conductor move constantly in random directions, like the particles of a gas. (More accurately, a Fermi gas.) To create a net flow of charge, the particles must also move together with an average drift rate. Electrons are the charge carriers in most metals and they follow an erratic path, bouncing from atom to atom, but generally drifting in the opposite direction of the electric field. The speed they drift at can be calculated from the equation:

$$I = nAvQ,$$

where

> I is the electric current.
>
> n is number of charged particles per unit volume (or charge carrier density).
>
> A is the cross-sectional area of the conductor.
>
> v is the drift velocity.
>
> Q is the charge on each particle.

Typically, electric charges in solids flow slowly. For example, in a copper wire of cross-section 0.5 mm^2, carrying a current of 5 A, the drift velocity of the electrons is on the order of a millimetre per second. To take a different example, in the near-vacuum inside a cathode ray tube, the electrons travel in near-straight lines at about a tenth of the speed of light.

Any accelerating electric charge, and therefore any changing electric current, gives rise to an electromagnetic wave that propagates at very high speed outside the surface of the conductor. This speed is usually a significant fraction of the speed of light, as can be deduced from Maxwell's Equations, and is therefore many times faster than the drift velocity of the electrons. For example, in AC power lines, the waves of electromagnetic energy propagate through the space between the wires, moving from a source to a distant load, even though the electrons in the wires only move back and forth over a tiny distance.

The ratio of the speed of the electromagnetic wave to the speed of light in free space is called the velocity factor, and depends on the electromagnetic properties of the conductor and the insulating materials surrounding it, and on their shape and size.

The magnitudes (not the natures) of these three velocities can be illustrated by an analogy with the three similar velocities associated with gases.

- The low drift velocity of charge carriers is analogous to air motion; in other words, winds.

- The high speed of electromagnetic waves is roughly analogous to the speed of sound in a gas (sound waves move through air much faster than large-scale motions such as convection).

- The random motion of charges is analogous to heat – the thermal velocity of randomly vibrating gas particles.

Direct Current

Direct current (DC) is the unidirectional flow of an electric charge. A battery is a prime example of DC power. Direct current may flow through a conductor such as a wire, but can also flow through semiconductors, insulators, or even through a vacuum as in electron or ion beams. The electric current flows in a constant direction, distinguishing it from alternating current (AC). A term formerly used for this type of current was galvanic current.

The abbreviations *AC* and *DC* are often used to mean simply *alternating* and *direct*, as when they modify *current* or *voltage*.

Direct current may be converted from an alternating current supply by use of a rectifier, which contains electronic elements (usually) or electromechanical elements (historically) that allow current to flow only in one direction. Direct current may be converted into alternating current via an inverter.

Direct current has many uses, from the charging of batteries to large power supplies for electronic systems, motors, and more. Very large quantities of direct-current power are used in production of aluminum and other electrochemical processes. It is also used for some railways, especially in urban areas. High-voltage direct current is used to transmit large amounts of power from remote generation sites or to interconnect alternating current power grids.

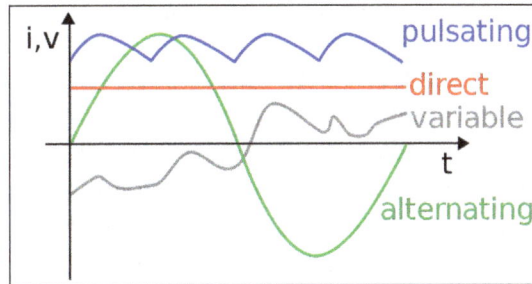

Direct Current (DC) (red line). The vertical axis shows current or voltage and the horizontal 't' axis measures time and shows the zero value.

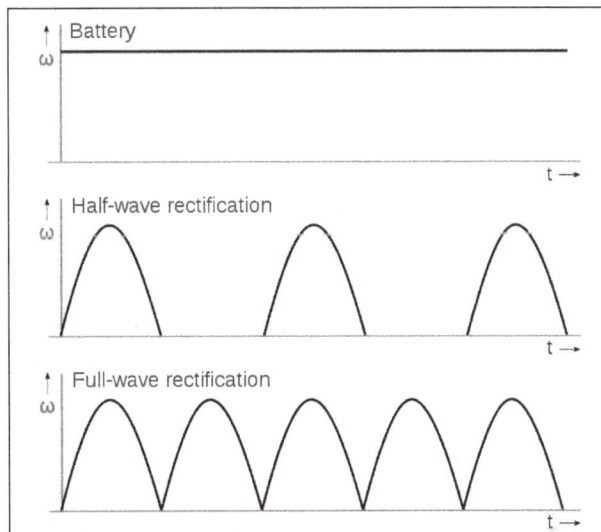

Types of direct current.

The term *DC* is used to refer to power systems that use only one polarity of voltage or current, and to refer to the constant, zero-frequency, or slowly varying local mean value of a voltage or current. For example, the voltage across a DC voltage source is constant as is the current through a DC current source. The DC solution of an electric circuit is the solution where all voltages and currents are constant. It can be shown that any stationary voltage or current waveform can be decomposed into a sum of a DC component and a zero-mean time-varying component; the DC component is defined to be the expected value, or the average value of the voltage or current over all time.

Although DC stands for "direct current", DC often refers to "constant polarity". Under this definition, DC voltages can vary in time, as seen in the raw output of a rectifier or the fluctuating voice signal on a telephone line.

Some forms of DC (such as that produced by a voltage regulator) have almost no variations in voltage, but may still have variations in output power and current.

Circuits

A direct current circuit is an electrical circuit that consists of any combination of constant voltage sources, constant current sources, and resistors. In this case, the circuit voltages and currents are independent of time. A particular circuit voltage or current does not depend on the past value of any circuit voltage or current. This implies that the system of equations that represent a DC circuit do not involve integrals or derivatives with respect to time.

If a capacitor or inductor is added to a DC circuit, the resulting circuit is not, strictly speaking, a DC circuit. However, most such circuits have a DC solution. This solution gives the circuit voltages and currents when the circuit is in DC steady state. Such a circuit is represented by a system of differential equations. The solution to these equations usually contain a time varying or transient part as well as constant or steady state part. It is this steady state part that is the DC solution. There are some circuits that do not have a DC solution. Two simple examples are a constant current source connected to a capacitor and a constant voltage source connected to an inductor.

In electronics, it is common to refer to a circuit that is powered by a DC voltage source such as a battery or the output of a DC power supply as a DC circuit even though what is meant is that the circuit is DC powered.

Alternating Current

Alternating current (AC) is an electric current which periodically reverses direction, in contrast to direct current (DC) which flows only in one direction. Alternating current is the form in which electric power is delivered to businesses and residences, and it is the form of electrical energy that consumers typically use when they plug kitchen appliances, televisions, fans and electric lamps into a wall socket. A common source of DC power is a battery cell in a flashlight. The abbreviations *AC* and *DC* are often used to mean simply *alternating* and *direct*, as when they modify *current* or *voltage*.

The usual waveform of alternating current in most electric power circuits is a sine wave, whose positive half-period corresponds with positive direction of the current and vice versa. In certain applications, like guitar amplifiers, different waveforms are used, such as triangular or square waves. Audio and radio signals carried on electrical wires are also examples of alternating current. These types of alternating current carry information such as sound (audio) or images (video) sometimes carried by modulation of an AC carrier signal. These currents typically alternate at higher frequencies than those used in power transmission.

Transmission, Distribution and Domestic Power Supply

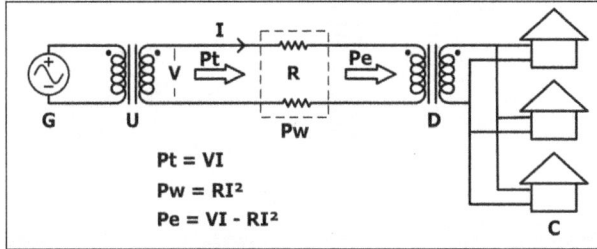

$$Pt = VI$$
$$Pw = RI^2$$
$$Pe = VI - RI^2$$

A schematic representation of long distance electric power transmission. C=consumers, D=step down transformer, G=generator, I=current in the wires, Pe=power reaching the end of the transmission line, Pt=power entering the transmission line, Pw=power lost in the transmission line, R=total resistance in the wires, V=voltage at the beginning of the transmission line, U=step up transformer.

Electrical energy is distributed as alternating current because AC voltage may be increased or decreased with a transformer. This allows the power to be transmitted through power lines efficiently at high voltage, which reduces the energy lost as heat due to resistance of the wire, and transformed to a lower, safer, voltage for use. Use of a higher voltage leads to significantly more efficient transmission of power. The power losses (P_w) in the wire are a product of the square of the current (I) and the resistance (R) of the wire, described by the formula:

$$P_w = I^2 R.$$

This means that when transmitting a fixed power on a given wire, if the current is halved (i.e. the voltage is doubled), the power loss due to the wire's resistance will be reduced to one quarter.

The power transmitted is equal to the product of the current and the voltage (assuming no phase difference); that is,

$$P_t = IV.$$

Consequently, power transmitted at a higher voltage requires less loss-producing current than for the same power at a lower voltage. Power is often transmitted at hundreds of kilovolts, and transformed to 100 V – 240 V for domestic use.

High voltages have disadvantages, such as the increased insulation required, and generally increased difficulty in their safe handling. In a power plant, energy is generated at a convenient voltage for the design of a generator, and then stepped up to a high voltage for transmission. Near the loads, the transmission voltage is stepped down to the voltages used by equipment. Consumer voltages vary somewhat depending on the country and size of load, but generally motors and lighting are built to use up to a few hundred volts between phases. The voltage delivered to equipment such as lighting and motor loads is standardized, with an allowable range of voltage over which equipment is expected to operate. Standard power utilization voltages and percentage tolerance vary in the different mains power systems found in the world. High-voltage direct-current (HVDC) electric power transmission systems have become more viable as technology has provided efficient means of changing the voltage of DC power. Transmission with high voltage direct current was not feasible in the early days of electric power transmission, as there was then no economically viable way to step down the voltage of DC for end user applications such as lighting incandescent bulbs.

High voltage transmission lines deliver power from electric generation plants over long distances using alternating current. These lines are located in eastern Utah.

Three-phase electrical generation is very common. The simplest way is to use three separate coils in the generator stator, physically offset by an angle of 120° (one-third of a complete 360° phase) to each other. Three current waveforms are produced that are equal in magnitude and 120° out of phase to each other. If coils are added opposite to these (60° spacing), they generate the same phases with reverse polarity and so can be simply wired together. In practice, higher "pole orders" are commonly used. For example, a 12-pole machine would have 36 coils (10° spacing). The advantage is that lower rotational speeds can be used to generate the same frequency. For example, a 2-pole machine running at 3600 rpm and a 12-pole machine running at 600 rpm produce the same frequency; the lower speed is preferable for larger machines. If the load on a three-phase system is balanced equally among the phases, no current flows through the neutral point. Even in the worst-case unbalanced (linear) load, the neutral current will not exceed the highest of the phase currents. Non-linear loads (e.g. the switch-mode power supplies widely used) may require an oversized neutral bus and neutral conductor in the upstream distribution panel to handle harmonics. Harmonics can cause neutral conductor current levels to exceed that of one or all phase conductors.

For three-phase at utilization voltages a four-wire system is often used. When stepping down three-phase, a transformer with a Delta (3-wire) primary and a Star (4-wire, center-earthed) secondary is often used so there is no need for a neutral on the supply side. For smaller customers (just how small varies by country and age of the installation) only a single phase and neutral, or two phases and neutral, are taken to the property. For larger installations all three phases and neutral are taken to the main distribution panel. From the three-phase main panel, both single and three-phase circuits may lead off. Three-wire single-phase systems, with a single center-tapped transformer giving two live conductors, is a common distribution scheme for residential and small commercial buildings in North America. This arrangement is sometimes incorrectly referred to as "two phase". A similar method is used for a different reason on construction sites in the UK. Small power tools and lighting are supposed to be supplied by a local center-tapped transformer with a voltage of 55 V between each power conductor and earth. This significantly reduces the risk of electric shock in the event that one of the live conductors becomes exposed through an equipment fault whilst still allowing a reasonable voltage of 110 V between the two conductors for running the tools.

A third wire, called the bond (or earth) wire, is often connected between non-current-carrying metal enclosures and earth ground. This conductor provides protection from electric shock due to accidental contact of circuit conductors with the metal chassis of portable appliances and tools. Bonding all non-current-carrying metal parts into one complete system ensures there is always a low electrical impedance path to ground sufficient to carry any fault current for as long as it takes for the system to clear the fault. This low impedance path allows the maximum amount of fault current, causing the overcurrent protection device (breakers, fuses) to trip or burn out as quickly as possible, bringing the electrical system to a safe state. All bond wires are bonded to ground at the main service panel, as is the neutral/identified conductor if present.

AC Power Supply Frequencies

The frequency of the electrical system varies by country and sometimes within a country; most electric power is generated at either 50 or 60 Hertz. Some countries have a mixture of 50 Hz and 60 Hz supplies, notably electricity power transmission in Japan. A low frequency eases the design of electric motors, particularly for hoisting, crushing and rolling applications, and commutator-type traction motors for applications such as railways. However, low frequency also causes noticeable flicker in arc lamps and incandescent light bulbs. The use of lower frequencies also provided the advantage of lower impedance losses, which are proportional to frequency. The original Niagara Falls generators were built to produce 25 Hz power, as a compromise between low frequency for traction and heavy induction motors, while still allowing incandescent lighting to operate (although with noticeable flicker). Most of the 25 Hz residential and commercial customers for Niagara Falls power were converted to 60 Hz by the late 1950s, although some 25cHz industrial customers still existed as of the start of the 21st century. 16.7 Hz power (formerly 16 2/3 Hz) is still used in some European rail systems, such as in Austria, Germany, Norway, Sweden and Switzerland. Off-shore, military, textile industry, marine, aircraft, and spacecraft applications sometimes use 400 Hz, for benefits of reduced weight of apparatus or higher motor speeds. Computer mainframe systems were often powered by 400 Hz or 415 Hz for benefits of ripple reduction while using smaller internal AC to DC conversion units. In any case, the input to the M-G set is the local customary voltage and frequency, variously 200 V (Japan), 208 V, 240 V (North America), 380 V, 400 V or 415 V (Europe), and variously 50 Hz or 60 Hz.

Effects at High Frequencies

A direct current flows uniformly throughout the cross-section of a uniform wire. An alternating current of any frequency is forced away from the wire's center, toward its outer surface. This is because the acceleration of an electric charge in an alternating current produces waves of electromagnetic radiation that cancel the propagation of electricity toward the center of materials with high conductivity. This phenomenon is called skin effect. At very high frequencies the current no longer flows *in* the wire, but effectively flows *on* the surface of the wire, within a thickness of a few skin depths. The skin depth is the thickness at which the current density is reduced by 63%. Even at relatively low frequencies used for power transmission (50 Hz – 60 Hz), non-uniform distribution of current still occurs in sufficiently thick conductors. For example, the skin depth of a copper conductor is approximately 8.57 mm at 60 Hz, so high current conductors are usually hollow to reduce their mass and cost. Since the current tends to flow in the periphery of conductors, the effective cross-section of the conductor is reduced. This increases the effective AC resistance of the conductor, since resistance is inversely proportional to the cross-sectional area. The AC resistance often is many times higher than the DC resistance, causing a much higher energy loss due to ohmic heating (also called I^2R loss).

Techniques for Reducing AC Resistance

For low to medium frequencies, conductors can be divided into stranded wires, each insulated from one another, with the relative positions of individual strands specially arranged within the conductor bundle. Wire constructed using this technique is called Litz wire. This measure helps to partially mitigate skin effect by forcing more equal current throughout the total cross section of the stranded conductors. Litz wire is used for making high-Q inductors, reducing losses in flexible conductors carrying very high currents at lower frequencies, and in the windings of devices carrying higher radio frequency current (up to hundreds of kilohertz), such as switch-mode power supplies and radio frequency transformers.

Techniques for Reducing Radiation Loss

As written above, an alternating current is made of electric charge under periodic acceleration, which causes radiation of electromagnetic waves. Energy that is radiated is lost. Depending on the frequency, different techniques are used to minimize the loss due to radiation.

Twisted Pairs

At frequencies up to about 1 GHz, pairs of wires are twisted together in a cable, forming a twisted pair. This reduces losses from electromagnetic radiation and inductive coupling. A twisted pair must be used with a balanced signalling system, so that the two wires carry equal but opposite currents. Each wire in a twisted pair radiates a signal, but it is effectively cancelled by radiation from the other wire, resulting in almost no radiation loss.

Coaxial Cables

Coaxial cables are commonly used at audio frequencies and above for convenience. A coaxial cable has a conductive wire inside a conductive tube, separated by a dielectric layer. The current flowing on the surface of the inner conductor is equal and opposite to the current flowing on the inner surface of the outer tube. The electromagnetic field is thus completely contained within the tube, and (ideally) no energy is lost to radiation or coupling outside the tube. Coaxial cables have acceptably small losses for frequencies up to about 5 GHz. For microwave frequencies greater than 5 GHz, the losses (due mainly to the electrical resistance of the central conductor) become too large, making waveguides a more efficient medium for transmitting energy. Coaxial cables with an air rather than solid dielectric are preferred as they transmit power with lower loss.

Waveguides

Waveguides are similar to coaxial cables, as both consist of tubes, with the biggest difference being that the waveguide has no inner conductor. Waveguides can have any arbitrary cross section, but rectangular cross sections are the most common. Because waveguides do not have an inner conductor to carry a return current, waveguides cannot deliver energy by means of an electric current, but rather by means of a *guided* electromagnetic field. Although surface currents do flow on the inner walls of the waveguides, those surface currents do not carry power. Power is carried by the guided electromagnetic fields. The surface currents are set up by the guided electromagnetic fields and have the effect of keeping the fields inside the waveguide and preventing leakage of the fields to the space outside the waveguide. Waveguides have dimensions comparable to the wavelength of the alternating current to be transmitted, so they are only feasible at microwave frequencies. In addition to this mechanical feasibility, electrical resistance of the non-ideal metals forming the walls of the waveguide cause dissipation of power (surface currents flowing on lossy conductors dissipate power). At higher frequencies, the power lost to this dissipation becomes unacceptably large.

Fiber Optics

At frequencies greater than 200 GHz, waveguide dimensions become impractically small, and the ohmic losses in the waveguide walls become large. Instead, fiber optics, which are a form of dielectric waveguides, can be used. For such frequencies, the concepts of voltages and currents are no longer used.

Mathematics of AC Voltages

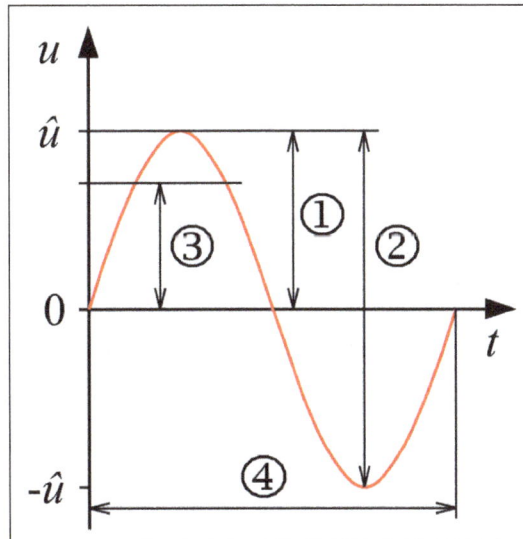

A sinusoidal alternating voltage: 1. Peak, also amplitude, 2. Peak-to-peak, 3. Effective value, 4. Period

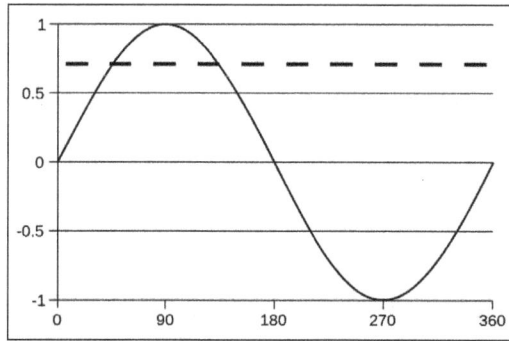

A sine wave, over one cycle (360°). The dashed line represents the root mean square (RMS) value at about 0.707.

Alternating currents are accompanied (or caused) by alternating voltages. An AC voltage v can be described mathematically as a function of time by the following equation:

$$v(t) = V_{peak} \sin(\omega t),$$

where,

- V_{peak} is the peak voltage (unit: volt).

- ω is the angular frequency (unit: radians per second).

  The angular frequency is related to the physical frequency, f (unit: hertz), which represents the number of cycles per second, by the equation $\omega = 2\pi f$.

- t is the time (unit: second).

The peak-to-peak value of an AC voltage is defined as the difference between its positive peak and its negative peak. Since the maximum value of $\sin(x)$ is +1 and the minimum value is −1, an AC voltage swings between $+V_{peak}$ and $-V_{peak}$. The peak-to-peak voltage, usually written as V_{pp} or V_{P-P}, is therefore $V_{peak} - (-V_{peak}) = 2V_{peak}$.

Power

The relationship between voltage and the power delivered is:

$$p(t) = \frac{v^2(t)}{R}$$

where R represents a load resistance.

Rather than using instantaneous power, $p(t)$, it is more practical to use a time averaged power (where the averaging is performed over any integer number of cycles). Therefore, AC voltage is often expressed as a root mean square (RMS) value, written as V_{rms}, because:

$$P_{time\ averaged} = \frac{V_{rms}^2}{R}.$$

Power Oscillation

$$v(t) = V_{peak} \sin(\omega t)$$

$$i(t) = \frac{v(t)}{R} = \frac{V_{peak}}{R} \sin(\omega t)$$

$$P(t) = v(t)i(t) = \frac{(V_{peak})^2}{R} \sin^2(\omega t)$$

Root Mean Square Voltage

Below it is assumed an AC waveform (with no DC component).

The RMS voltage is the square root of the mean over one cycle of the square of the instantaneous voltage.

- For an arbitrary periodic waveform $v(t)$ of period T:

$$V_{rms} = \sqrt{\frac{1}{T} \int_0^T [v(t)]^2 dt}.$$

- For a sinusoidal voltage:

$$V_{rms} = \sqrt{\frac{1}{T} \int_0^T [V_{pk} \sin(\omega t + \phi)]^2 dt}$$

$$= V_{pk} \sqrt{\frac{1}{2T} \int_0^T [1 - \cos(2\omega t + 2\phi)] dt}$$

$$= V_{pk} \sqrt{\frac{1}{2T} \int_0^T dt}$$

$$= \frac{V_{pk}}{\sqrt{2}}$$

where the trigonometric identity $\sin^2(x) = \frac{1 - \cos(2x)}{2}$ has been used and the factor $\sqrt{2}$ is called the crest factor, which varies for different waveforms.

- For a triangle waveform centered about zero,

$$V_{rms} = \frac{V_{peak}}{\sqrt{3}}.$$

- For a square waveform centered about zero,

$$V_{rms} = V_{peak}.$$

Example

To illustrate these concepts, consider a 230 V AC mains supply used in many countries around the world. It is so called because its root mean square value is 230 V. This means that the time-averaged power delivered is equivalent to the power delivered by a DC voltage of 230 V. To determine the peak voltage (amplitude), we can rearrange the above equation to:

$$V_{peak} = \sqrt{2}\, V_{rms}.$$

For 230 V AC, the peak voltage V_{peak} is therefore $230\text{ V} \times \sqrt{2}$, which is about 325 V. During the course of one cycle the voltage rises from zero to 325 V, falls through zero to −325 V, and returns to zero.

Voltage

Voltage, electric potential difference, electric pressure or electric tension is the difference in electric potential between two points. The difference in electric potential between two points (i.e., voltage) in a static electric field is defined as the work needed per unit of charge to move a test charge between the two points. In the International System of Units, the derived unit for voltage is named *volt*. In SI units, work per unit charge is expressed as joules per coulomb, where 1 volt = 1 joule (of work) per 1 coulomb (of charge). The official SI definition for *volt* uses power and current, where 1 volt = 1 watt (of power) per 1 ampere (of current). This definition is equivalent to the more commonly used 'joules per coulomb'. Voltage or electric potential difference is denoted symbolically by ΔV, but more often simply as V, for instance in the context of Ohm's or Kirchhoff's circuit laws.

Electric potential differences between points can be caused by electric charge, by electric current through a magnetic field, by time-varying magnetic fields, or some combination of these three. A voltmeter can be used to measure the voltage (or potential difference) between two points in a system; often a common reference potential such as the ground of the system is used as one of the points. A voltage may represent either a source of energy (electromotive force) or lost, used, or stored energy (potential drop).

There are multiple useful ways to define voltage, including the standard definition. There are also other useful definitions of work per charges.

Roughly speaking, voltage is defined so that negatively charged objects are pulled towards higher voltages, while positively charged objects are pulled towards lower voltages. Therefore, the conventional current in a wire or resistor always flows from higher voltage to lower voltage.

Historically, voltage has been referred to using terms like "tension" and "pressure". Even today, the term "tension" is still used, for example within the phrase "high tension" (HT) which is commonly used in thermionic valve (vacuum tube) based electronics.

Definition as Potential of Electric Field

The voltage increase from some point x_A to some point x_B is given by,

$$\Delta V_{AB} = V(x_B) - V(x_A)$$
$$= -\int_{r_0}^{x_B} \vec{E} \cdot d\vec{l} - \left(-\int_{r_0}^{x_A} \vec{E} \cdot d\vec{l} \right)$$
$$= -\int_{x_A}^{x_B} \vec{E} \cdot d\vec{l}$$

In this case, the voltage increase from point A to point B is equal to the work which would have to be done per unit charge, against the electric field, to move the charge from A to B without causing any acceleration. Mathematically, this is expressed as the line integral of the electric field along that path. Under this definition, the voltage difference between two points is not uniquely defined when there are time-varying magnetic fields since the electric force is not a conservative force in such cases.

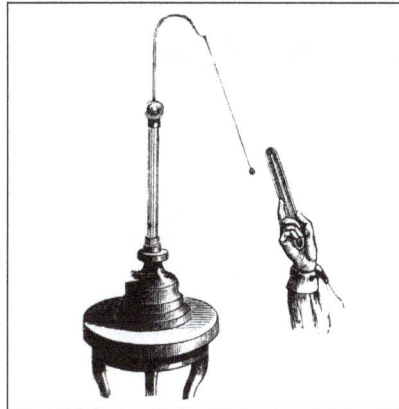

The electric field around the rod exerts a force on the charged pith ball, in an electroscope.

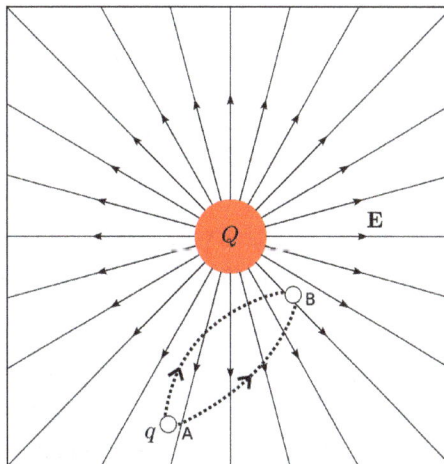

In a static field, the work is independent of the path.

If this definition of voltage is used, any circuit where there are time-varying magnetic fields, such as circuits containing inductors, will not have a well-defined voltage between nodes in the circuit.

However, if magnetic fields are suitably contained to each component, then the electric field is conservative in the region exterior to the components, and voltages are well-defined in that region. In this case, the voltage across an inductor, viewed externally, turns out to be,

$$\Delta V = -L \frac{dI}{dt}$$

despite the fact that, internally, the electric field in the coil is zero (assuming it is a perfect conductor).

Definition Via Decomposition of Electric Field

Using the above definition, the electric potential is not defined whenever magnetic fields change with time. In physics, it's sometimes useful to generalize the electric potential by only considering the conservative part of the electric field. This is done by the following decomposition used in electrodynamics:

$$\vec{E} = -\nabla V - \frac{\partial \vec{A}}{\partial t}$$

where $\vec{A}$ is the magnetic vector potential. The above decomposition is justified by Helmholtz's theorem.

In this case, the voltage increase from x_A to x_B is given by,

$$\Delta V_{AB} = -\int_{x_A}^{x_B} \vec{E}_{conservative} \cdot d\vec{l}$$

$$= -\int_{x_A}^{x_B} \left(\vec{E} + \frac{\partial \vec{A}}{\partial t} \right) \cdot d\vec{l}$$

$$= -\int_{x_A}^{x_B} (\vec{E} - \vec{E}_{induced}) \cdot d\vec{l}$$

where $\vec{E}_{induced}$ is the rotational electric field due to time-varying magnetic fields. In this case, the voltage between points is always uniquely defined.

Treatment in Circuit Theory

In circuit analysis and electrical engineering, the voltage across an inductor is not considered to be zero or undefined, as the standard definition would suggest. This is because electrical engineers use a lumped element model to represent and analyze circuits.

When using a lumped element model, it is assumed that there are no magnetic fields in the region surrounding the circuit and that the effects of these are contained in 'lumped elements', which are idealized and self-contained circuit elements used to model physical components. If the assumption of negligible leaked fields is too inaccurate, their effects can be modelled by parasitic components.

In the case of a physical inductor though, the ideal lumped representation is often accurate. This is because the leaked fields of the inductor are generally negligible, especially if the inductor is a toroid. If leaked fields are negligible, we find that:

$$\int_{exterior} \vec{E} \cdot d\vec{l} = -L\frac{dI}{dt}$$

is path-independent, and there is a well-defined voltage across the inductor's terminals. This is the reason that measurements with a voltmeter across an inductor are often reasonably independent of the placement of the test leads.

Volt

The volt (symbol: V) is the derived unit for electric potential, electric potential difference, and electromotive force. The volt is named in honour of the Italian physicist Alessandro Volta (1745–1827), who invented the voltaic pile, possibly the first chemical battery.

Hydraulic Analogy

A simple analogy for an electric circuit is water flowing in a closed circuit of pipework, driven by a mechanical pump. This can be called a "water circuit". Potential difference between two points corresponds to the pressure difference between two points. If the pump creates a pressure difference between two points, then water flowing from one point to the other will be able to do work, such as driving a turbine. Similarly, work can be done by an electric current driven by the potential difference provided by a battery. For example, the voltage provided by a sufficiently-charged automobile battery can "push" a large current through the windings of an automobile's starter motor. If the pump isn't working, it produces no pressure difference, and the turbine will not rotate. Likewise, if the automobile's battery is very weak or "dead" (or "flat"), then it will not turn the starter motor.

The hydraulic analogy is a useful way of understanding many electrical concepts. In such a system, the work done to move water is equal to the pressure multiplied by the volume of water moved. Similarly, in an electrical circuit, the work done to move electrons or other charge-carriers is equal to "electrical pressure" multiplied by the quantity of electrical charges moved. In relation to "flow", the larger the "pressure difference" between two points (potential difference or water pressure difference), the greater the flow between them (electric current or water flow).

Applications

Specifying a voltage measurement requires explicit or implicit specification of the points across which the voltage is measured. When using a voltmeter to measure potential difference, one electrical lead of the voltmeter must be connected to the first point, one to the second point.

A common use of the term "voltage" is in describing the voltage dropped across an electrical device (such as a resistor). The voltage drop across the device can be understood as the difference between measurements at each terminal of the device with respect to a common reference point (or ground). The voltage drop is the difference between the two readings. Two points in an electric circuit that are connected by an ideal conductor without resistance and not within a changing magnetic field have a voltage of zero. Any two points with the same potential may be connected by a conductor and no current will flow between them.

Addition of Voltages

The voltage between A and C is the sum of the voltage between A and B and the voltage between B and C. The various voltages in a circuit can be computed using Kirchhoff's circuit laws.

When talking about alternating current (AC) there is a difference between instantaneous voltage and average voltage. Instantaneous voltages can be added for direct current (DC) and AC, but average voltages can be meaningfully added only when they apply to signals that all have the same frequency and phase.

Measuring Instruments

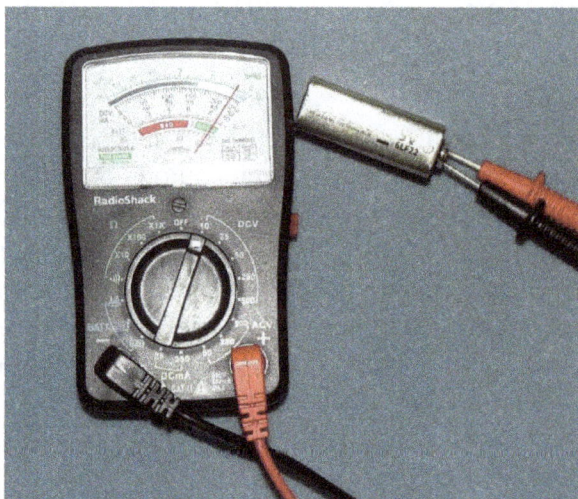

Multimeter set to measure voltage.

Instruments for measuring voltages include the voltmeter, the potentiometer, and the oscilloscope. Analog voltmeters, such as moving-coil instruments, work by measuring the current through a fixed resistor, which, according to Ohm's Law, is proportional to the voltage across the resistor. The potentiometer works by balancing the unknown voltage against a known voltage in a bridge circuit. The cathode-ray oscilloscope works by amplifying the voltage and using it to deflect an electron beam from a straight path, so that the deflection of the beam is proportional to the voltage.

Typical Voltages

A common voltage for flashlight batteries is 1.5 volts (DC). A common voltage for automobile batteries is 12 volts (DC).

Common voltages supplied by power companies to consumers are 110 to 120 volts (AC) and 220 to 240 volts (AC). The voltage in electric power transmission lines used to distribute electricity from power stations can be several hundred times greater than consumer voltages, typically 110 to 1200 kV (AC).

The voltage used in overhead lines to power railway locomotives is between 12 kV and 50 kV (AC) or between 1.5 kV and 3 kV (DC).

Galvani Potential vs. Electrochemical Potential

Inside a conductive material, the energy of an electron is affected not only by the average electric potential, but also by the specific thermal and atomic environment that it is in. When a voltmeter is connected between two different types of metal, it measures not the electrostatic potential difference, but instead something else that is affected by thermodynamics. The quantity measured by a voltmeter is the negative of the difference of the electrochemical potential of electrons (Fermi level) divided by the electron charge and commonly referred to as the voltage difference, while the pure unadjusted electrostatic potential (not measurable with a voltmeter) is sometimes called Galvani potential. The terms "voltage" and "electric potential" are ambiguous in that, in practice, they can refer to *either* of these in different contexts.

Ohm's Law

Ohm's law states that the current through a conductor between two points is directly proportional to the voltage across the two points. Introducing the constant of proportionality, the resistance, one arrives at the usual mathematical equation that describes this relationship:

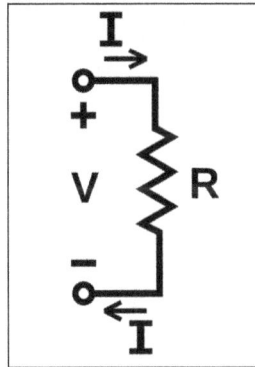

V, I, and R, the parameters of Ohm's law.

$$I = \frac{V}{R},$$

where I is the current through the conductor in units of amperes, V is the voltage measured *across* the conductor in units of volts, and R is the resistance of the conductor in units of ohms. More specifically, Ohm's law states that the R in this relation is constant, independent of the current. Ohm's law is an empirical relation which accurately describes the conductivity of the vast majority of electrically conductive materials over many orders of magnitude of current. However some materials do not obey Ohm's law, these are called non-ohmic.

The law was named after the German physicist Georg Ohm, who, in a treatise published in 1827, described measurements of applied voltage and current through simple electrical circuits containing various lengths of wire. Ohm explained his experimental results by a slightly more complex equation than the modern form above.

In physics, the term *Ohm's law* is also used to refer to various generalizations of the law; for example the vector form of the law used in electromagnetics and material science:

$$J = \sigma E,$$

where J is the current density at a given location in a resistive material, E is the electric field at that location, and σ (sigma) is a material-dependent parameter called the conductivity. This reformulation of Ohm's law is due to Gustav Kirchhoff.

Microscopic Origins

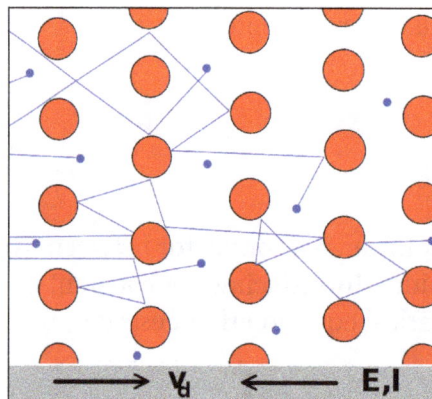

Drude Model electrons (shown here in blue) constantly bounce
among heavier, stationary crystal ions (shown in red).

The dependence of the current density on the applied electric field is essentially quantum mechanical in nature; A qualitative description leading to Ohm's law can be based upon classical mechanics using the Drude model developed by Paul Drude in 1900.

The Drude model treats electrons (or other charge carriers) like pinballs bouncing among the ions that make up the structure of the material. Electrons will be accelerated in the opposite direction to the electric field by the average electric field at their location. With each collision, though, the electron is deflected in a random direction with a velocity that is much larger than the velocity gained by the electric field. The net result is that electrons take a zigzag path due to the collisions, but generally drift in a direction opposing the electric field.

The drift velocity then determines the electric current density and its relationship to E and is independent of the collisions. Drude calculated the average drift velocity from $p = -eE\tau$ where p is the average momentum, $-e$ is the charge of the electron and τ is the average time between the collisions. Since both the momentum and the current density are proportional to the drift velocity, the current density becomes proportional to the applied electric field; this leads to Ohm's law.

Hydraulic Analogy

A hydraulic analogy is sometimes used to describe Ohm's law. Water pressure, measured by pascals (or PSI), is the analog of voltage because establishing a water pressure difference between two points along a (horizontal) pipe causes water to flow. Water flow rate, as in liters per second, is the analog of current, as in coulombs per second. Finally, flow restrictors—such as apertures placed in pipes between points where the water pressure is measured—are the analog of resistors. We say that the rate of water flow through an aperture restrictor is proportional to the difference in water pressure across the restrictor. Similarly, the rate of flow of electrical charge, that is, the electric current, through an electrical resistor is proportional to the difference in voltage measured across the resistor.

Flow and pressure variables can be calculated in fluid flow network with the use of the hydraulic ohm analogy. The method can be applied to both steady and transient flow situations. In the linear laminar flow region, Poiseuille's law describes the hydraulic resistance of a pipe, but in the turbulent flow region the pressure–flow relations become nonlinear.

The hydraulic analogy to Ohm's law has been used, for example, to approximate blood flow through the circulatory system.

Circuit Analysis

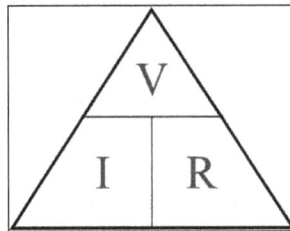

Ohm's law triangle.

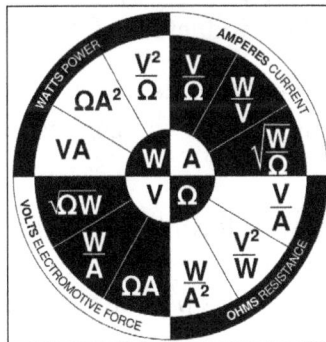

Ohm's law wheel with international unit symbols.

In circuit analysis, three equivalent expressions of Ohm's law are used interchangeably:

$$I = \frac{V}{R} \quad \text{or} \quad V = IR \quad \text{or} \quad R = \frac{V}{I}.$$

Each equation is quoted by some sources as the defining relationship of Ohm's law, or all three are quoted, or derived from a proportional form, or even just the two that do not correspond to Ohm's original statement may sometimes be given.

The interchangeability of the equation may be represented by a triangle, where V (voltage) is placed on the top section, the I (current) is placed to the left section, and the R (resistance) is placed to the right. The line that divides the left and right sections indicates multiplication, and the divider between the top and bottom sections indicates division (hence the division bar).

Resistive Circuits

Resistors are circuit elements that impede the passage of electric charge in agreement with Ohm's law, and are designed to have a specific resistance value R. In a schematic diagram the resistor is shown as a zig-zag symbol. An element (resistor or conductor) that behaves according to Ohm's law over some operating range is referred to as an *ohmic device* (or an *ohmic resistor*) because Ohm's law and a single value for the resistance suffice to describe the behavior of the device over that range.

Ohm's law holds for circuits containing only resistive elements (no capacitances or inductances) for all forms of driving voltage or current, regardless of whether the driving voltage or current is constant (DC) or time-varying such as AC. At any instant of time Ohm's law is valid for such circuits.

Resistors which are in *series* or in *parallel* may be grouped together into a single "equivalent resistance" in order to apply Ohm's law in analyzing the circuit.

Reactive Circuits with Time-varying Signals

When reactive elements such as capacitors, inductors, or transmission lines are involved in a circuit to which AC or time-varying voltage or current is applied, the relationship between voltage and current becomes the solution to a differential equation, so Ohm's law (as defined above) does not directly apply since that form contains only resistances having value R, not complex impedances which may contain capacitance ("C") or inductance ("L").

Equations for time-invariant AC circuits take the same form as Ohm's law. However, the variables are generalized to complex numbers and the current and voltage waveforms are complex exponentials.

In this approach, a voltage or current waveform takes the form Ae^{st}, where t is time, s is a complex parameter, and A is a complex scalar. In any linear time-invariant system, all of the currents and voltages can be expressed with the same s parameter as the input to the system, allowing the time-varying complex exponential term to be canceled out and the system described algebraically in terms of the complex scalars in the current and voltage waveforms.

The complex generalization of resistance is impedance, usually denoted Z; it can be shown that for an inductor,

$$Z = sL$$

and for a capacitor,

$$Z = \frac{1}{sC}.$$

We can now write,

$$V = I \cdot Z$$

where V and I are the complex scalars in the voltage and current respectively and Z is the complex impedance.

This form of Ohm's law, with Z taking the place of R, generalizes the simpler form. When Z is complex, only the real part is responsible for dissipating heat.

In the general AC circuit, Z varies strongly with the frequency parameter s, and so also will the relationship between voltage and current.

For the common case of a steady sinusoid, the s parameter is taken to be $j\omega$, corresponding to a complex sinusoid $Ae^{j\omega t}$. The real parts of such complex current and voltage waveforms describe the actual sinusoidal currents and voltages in a circuit, which can be in different phases due to the different complex scalars.

Linear Approximations

Ohm's law is one of the basic equations used in the analysis of electrical circuits. It applies to both metal conductors and circuit components (resistors) specifically made for this behaviour. Both are ubiquitous in electrical engineering. Materials and components that obey Ohm's law are described as "ohmic" which means they produce the same value for resistance (R = V/I) regardless of the value of V or I which is applied and whether the applied voltage or current is DC (direct current) of either positive or negative polarity or AC (alternating current).

In a true ohmic device, the same value of resistance will be calculated from R = V/I regardless of the value of the applied voltage V. That is, the ratio of V/I is constant, and when current is plotted as a function of voltage the curve is *linear* (a straight line). If voltage is forced to some value V, then that voltage V divided by measured current I will equal R. Or if the current is forced to some value I, then the measured voltage V divided by that current I is also R. Since the plot of I versus V is a straight line, then it is also true that for any set of two different voltages V_1 and V_2 applied across a given device of resistance R, producing currents $I_1 = V_1/R$ and $I_2 = V_2/R$, that the ratio $(V_1-V_2)/(I_1-I_2)$ is also a constant equal to R. The operator "delta" (Δ) is used to represent a difference in a quantity, so we can write $\Delta V = V_1-V_2$ and $\Delta I = I_1-I_2$. Summarizing, for any truly ohmic device having resistance R, $V/I = \Delta V/\Delta I = R$ for any applied voltage or current or for the difference between any set of applied voltages or currents.

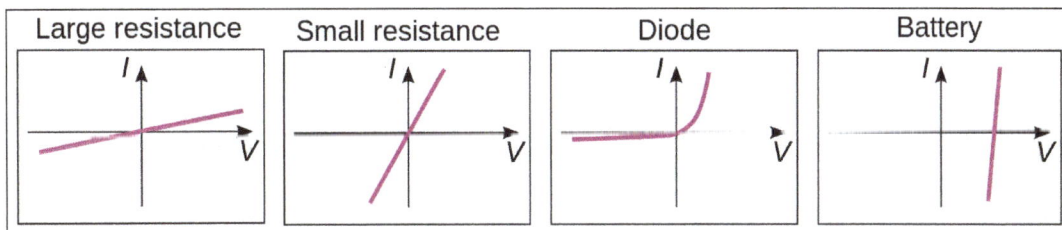

The I–V curves of four devices: Two resistors, a diode, and a battery. The two resistors follow Ohm's law: The plot is a straight line through the origin. The other two devices do *not* follow Ohm's law.

There are, however, components of electrical circuits which do not obey Ohm's law; that is, their relationship between current and voltage (their I–V curve) is *nonlinear* (or non-ohmic). An example is the p-n junction diode (curve at right). As seen in the figure, the current does not increase linearly with applied voltage for a diode. One can determine a value of current (I) for a given value of applied voltage (V) from the curve, but not from Ohm's law, since the value of "resistance" is not constant as a function of applied voltage. Further, the current only increases significantly if the applied voltage is positive, not negative. The ratio V/I for some point along the nonlinear curve is sometimes called the *static*, or *chordal*, or DC, resistance, but as seen in the figure the value of total V over total I varies depending on the particular point along the nonlinear curve which is chosen. This means the "DC resistance" V/I at some point on the curve is not the same as what would be determined by applying an AC signal having peak amplitude ΔV volts or ΔI amps centered at that same point along the curve and measuring $\Delta V/\Delta I$. However, in some diode applications, the AC signal applied to the device is small and it is possible to analyze the circuit in terms of the *dynamic*, *small-signal*, or *incremental* resistance, defined as the one over the slope of the V–I curve at the average value (DC operating point) of the voltage (that is, one over the derivative of current with respect to voltage). For sufficiently small signals, the dynamic resistance allows the Ohm's law small signal resistance to be calculated as approximately one over the slope of a line drawn tangentially to the V-I curve at the DC operating point.

Temperature Effects

Ohm's law has sometimes been stated as, "for a conductor in a given state, the electromotive force is proportional to the current produced." That is, that the resistance, the ratio of the applied electromotive force (or voltage) to the current, "does not vary with the current strength." The qualifier "in a given state" is usually interpreted as meaning "at a constant temperature," since the resistivity of materials is usually temperature dependent. Because the conduction of current is related to Joule heating of the conducting body, according to Joule's first law, the temperature of a conducting body may change when it carries a current. The dependence of resistance on temperature therefore makes resistance depend upon the current in a typical experimental setup, making the law in this form difficult to directly verify. Maxwell and others worked out several methods to test the law experimentally in 1876, controlling for heating effects.

Relation to Heat Conductions

Ohm's principle predicts the flow of electrical charge (i.e. current) in electrical conductors when subjected to the influence of voltage differences; Jean-Baptiste-Joseph Fourier's principle predicts the flow of heat in heat conductors when subjected to the influence of temperature differences.

The same equation describes both phenomena, the equation's variables taking on different meanings in the two cases. Specifically, solving a heat conduction (Fourier) problem with *temperature* (the driving "force") and *flux of heat* (the rate of flow of the driven "quantity", i.e. heat energy) variables also solves an analogous electrical conduction (Ohm) problem having *electric potential* (the driving "force") and *electric current* (the rate of flow of the driven "quantity", i.e. charge) variables.

The basis of Fourier's work was his clear conception and definition of thermal conductivity. He assumed that, all else being the same, the flux of heat is strictly proportional to the gradient of temperature. Although undoubtedly true for small temperature gradients, strictly proportional behavior will be lost when real materials (e.g. ones having a thermal conductivity that is a function of temperature) are subjected to large temperature gradients.

A similar assumption is made in the statement of Ohm's law: other things being alike, the strength of the current at each point is proportional to the gradient of electric potential. The accuracy of the assumption that flow is proportional to the gradient is more readily tested, using modern measurement methods, for the electrical case than for the heat case.

Electrical Resistance and Conductance

The electrical resistance of an object is a measure of its opposition to the flow of electric current. The inverse quantity is electrical conductance, and is the ease with which an electric current passes. Electrical resistance shares some conceptual parallels with the notion of mechanical friction. The SI unit of electrical resistance is the ohm (Ω), while electrical conductance is measured in siemens (S).

The resistance of an object depends in large part on the material it is made of—objects made of electrical insulators like rubber tend to have very high resistance and low conductivity, while objects made of electrical conductors like metals tend to have very low resistance and high conductivity. This material dependence is quantified by resistivity or conductivity. However, resistance and conductance are extensive rather than bulk properties, meaning that they also depend on the size and shape of an object. For example, a wire's resistance is higher if it is long and thin, and lower if it is short and thick. All objects show some resistance, except for superconductors, which have a resistance of zero.

The resistance (R) of an object is defined as the ratio of voltage across it (V) to current through it (I), while the conductance (G) is the inverse:

$$R = \frac{V}{I}, \qquad G = \frac{I}{V} = \frac{1}{R}$$

For a wide variety of materials and conditions, V and I are directly proportional to each other, and therefore R and G are constants (although they will depend on the size and shape of the object, the material it is made of, and other factors like temperature or strain). This proportionality is called Ohm's law, and materials that satisfy it are called *ohmic* materials.

In other cases, such as a transformer, diode or battery, V and I are *not* directly proportional. The ratio V over I is sometimes still useful, and is referred to as a "chordal resistance" or "static resistance", since it corresponds to the inverse slope of a chord between the origin and an $I–V$ curve. In other situations, the derivative $\frac{dV}{dI}$ may be most useful; this is called the "differential resistance".

The hydraulic analogy compares electric current flowing through circuits to water flowing through pipes. When a pipe (left) is filled with hair (right), it takes a larger pressure to achieve the same flow of water. Pushing electric current through a large resistance is like pushing water through a pipe clogged with hair: It requires a larger push (electromotive force) to drive the same flow (electric current).

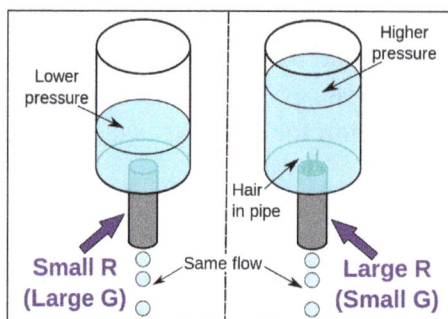

In the hydraulic analogy, current flowing through a wire (or resistor) is like water flowing through a pipe, and the voltage drop across the wire is like the pressure drop that pushes water through the pipe. Conductance is proportional to how much flow occurs for a given pressure, and resistance is proportional to how much pressure is required to achieve a given flow. (Conductance and resistance are reciprocals.)

The voltage drop (i.e., difference between voltages on one side of the resistor and the other), not the voltage itself, provides the driving force pushing current through a resistor. In hydraulics, it is similar: The pressure difference between two sides of a pipe, not the pressure itself, determines the flow through it. For example, there may be a large water pressure above the pipe, which tries to push water down through the pipe. But there may be an equally large water pressure below the pipe, which tries to push water back up through the pipe. If these pressures are equal, no water flows. (In the image at right, the water pressure below the pipe is zero.)

The resistance and conductance of a wire, resistor, or other element is mostly determined by two properties:

- Geometry (shape)

- Material

Geometry is important because it is more difficult to push water through a long, narrow pipe than a wide, short pipe. In the same way, a long, thin copper wire has higher resistance (lower conductance) than a short, thick copper wire.

Materials are important as well. A pipe filled with hair restricts the flow of water more than a clean pipe of the same shape and size. Similarly, electrons can flow freely and easily through a copper wire, but cannot flow as easily through a steel wire of the same shape and size, and they essentially cannot flow at all through an insulator like rubber, regardless of its shape. The difference between copper, steel, and rubber is related to their microscopic structure and electron configuration, and is quantified by a property called resistivity.

In addition to geometry and material, there are various other factors that influence resistance and conductance, such as temperature.

Conductors and Resistors

Substances in which electricity can flow are called conductors. A piece of conducting material of a particular resistance meant for use in a circuit is called a resistor. Conductors are made of high-conductivity materials such as metals, in particular copper and aluminium. Resistors, on the other hand, are made of a wide variety of materials depending on factors such as the desired resistance, amount of energy that it needs to dissipate, precision, and costs.

A 65 Ω resistor, as identified by its electronic color code (blue–green–black-gold-red).
An ohmmeter could be used to verify this value.

Relation to Resistivity and Conductivity

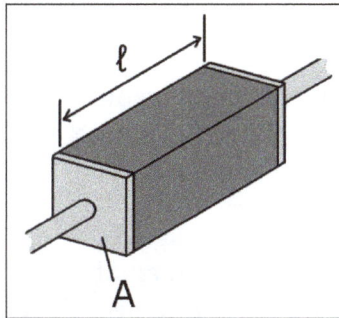

A piece of resistive material with electrical contacts on both ends.

The resistance of a given object depends primarily on two factors: What material it is made of, and its shape. For a given material, the resistance is inversely proportional to the cross-sectional area; for example, a thick copper wire has lower resistance than an otherwise-identical thin copper wire. Also, for a given material, the resistance is proportional to the length; for example, a long copper wire has higher resistance than an otherwise-identical short copper wire. The resistance R and conductance G of a conductor of uniform cross section, therefore, can be computed as,

$$R = \rho \frac{\ell}{A},$$

$$G = \sigma \frac{A}{\ell}.$$

where ℓ is the length of the conductor, measured in metres (m), A is the cross-sectional area of the conductor measured in square metres (m²), σ (sigma) is the electrical conductivity measured in siemens per meter (S·m⁻¹), and ρ (rho) is the electrical resistivity (also called *specific electrical resistance*) of the material, measured in ohm-metres (Ω·m). The resistivity and conductivity are proportionality constants, and therefore depend only on the material the wire is made of, not the geometry of the wire. Resistivity and conductivity are reciprocals: $\rho = 1/\sigma$. Resistivity is a measure of the material's ability to oppose electric current.

This formula is not exact, as it assumes the current density is totally uniform in the conductor, which is not always true in practical situations. However, this formula still provides a good approximation for long thin conductors such as wires.

Another situation for which this formula is not exact is with alternating current (AC), because the skin effect inhibits current flow near the center of the conductor. For this reason, the *geometrical* cross-section is different from the *effective* cross-section in which current actually flows, so resistance is higher than expected. Similarly, if two conductors near each other carry AC current, their resistances increase due to the proximity effect. At commercial power frequency, these effects are significant for large conductors carrying large currents, such as busbars in an electrical substation, or large power cables carrying more than a few hundred amperes.

The resistivity of different materials varies by an enormous amount: For example, the conductivity of teflon is about 10^{30} times lower than the conductivity of copper. Why is there such a difference? Loosely speaking, a metal has large numbers of "delocalized" electrons that are not stuck in any one place, but free to move across large distances, whereas in an insulator (like teflon), each electron is tightly bound to a single molecule, and a great force is required to pull it away. Semiconductors lie between these two extremes.

Resistivity varies with temperature. In semiconductors, resistivity also changes when exposed to light.

Measuring Resistance

An instrument for measuring resistance is called an ohmmeter. Simple ohmmeters cannot measure low resistances accurately because the resistance of their measuring leads causes a voltage drop that interferes with the measurement, so more accurate devices use four-terminal sensing.

Typical Resistances

Component	Resistance (Ω)
1 meter of copper wire with 1 mm diameter	0.02
1 km overhead power line (*typical*)	0.03
AA battery (*typical internal resistance*)	0.1
Incandescent light bulb filament (*typical*)	200–1000
Human body	1000 to 100,000

Static and Differential Resistance

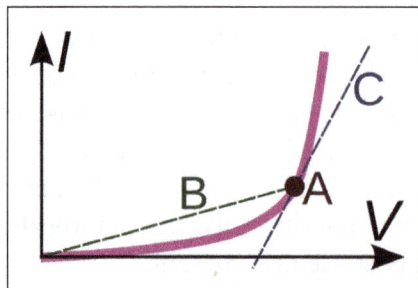

The IV curve of a non-ohmic device (purple). The static Resistance at point *A* is the inverse slope of line *B* through the origin. The differential Resistance at *A* is the inverse slope of tangent line *C*.

Many electrical elements, such as diodes and batteries do *not* satisfy Ohm's law. These are called *non-ohmic* or *non-linear*, and their *I–V* curves are *not* straight lines through the origin.

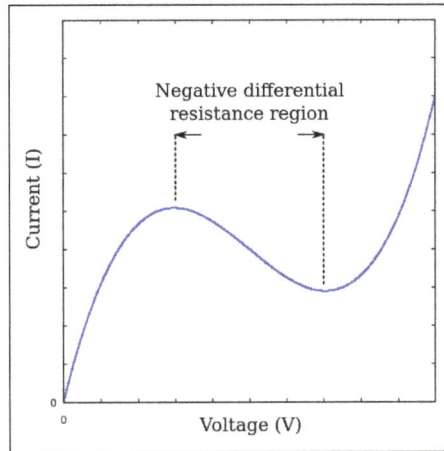

The IV curve of a component with negative differential resistance, an
unusual phenomenon where the IV curve is non-monotonic.

Resistance and conductance can still be defined for non-ohmic elements. However, unlike ohmic
resistance, non-linear resistance is not constant but varies with the voltage or current through the
device; i.e., its operating point. There are two types of resistance:

- Static resistance (also called *chordal* or *DC resistance*) – This corresponds to the usual
  definition of resistance; the voltage divided by the current

$$R_{static} = \frac{V}{I}.$$

- It is the slope of the line (chord) from the origin through the point on the curve. Static resistance determines the power dissipation in an electrical component. Points on the IV curve
  located in the 2nd or 4th quadrants, for which the slope of the chordal line is negative, have
  negative static resistance. Passive devices, which have no source of energy, cannot have negative static resistance. However active devices such as transistors or op-amps can synthesize
  negative static resistance with feedback, and it is used in some circuits such as gyrators.

$$R_{diff} = \frac{dV}{dI}.$$

Differential resistance (also called dynamic, incremental or small signal resistance) – Differential resistance is the derivative of the voltage with respect to the current; the slope of the IV curve at a point

AC Circuits

Impedance and Admittance

The voltage (red) and current (blue) versus time (horizontal axis) for a capacitor (top) and inductor (bottom). Since the amplitude of the current and voltage sinusoids are the same, the absolute
value of impedance is 1 for both the capacitor and the inductor (in whatever units the graph is
using). On the other hand, the phase difference between current and voltage is −90° for the capacitor; therefore, the complex phase of the impedance of the capacitor is −90°. Similarly, the phase
difference between current and voltage is +90° for the inductor; therefore, the complex phase of
the impedance of the inductor is +90°.

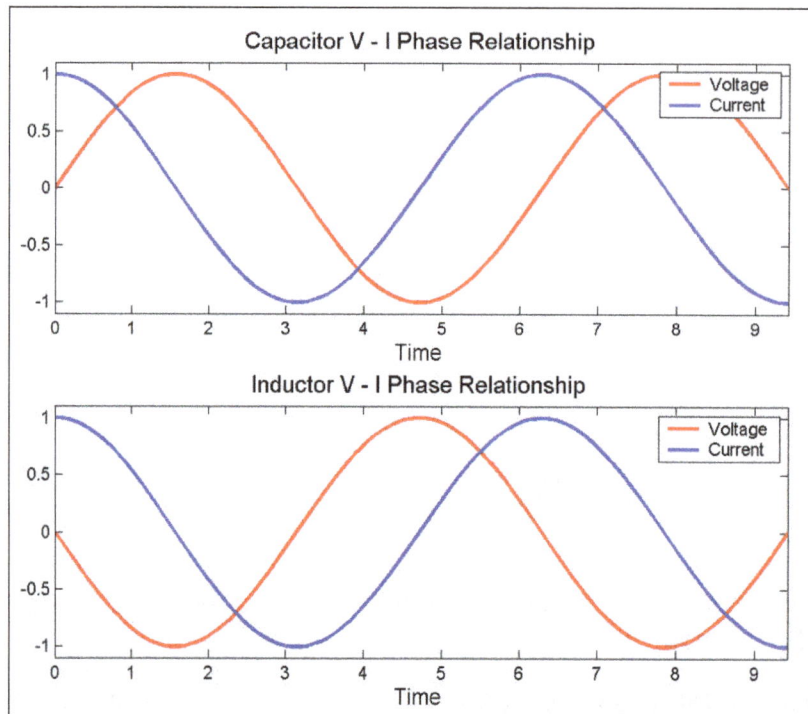

Capacitor V - I Phase Relationship

Inductor V - I Phase Relationship

When an alternating current flows through a circuit, the relation between current and voltage across a circuit element is characterized not only by the ratio of their magnitudes, but also the difference in their phases. For example, in an ideal resistor, the moment when the voltage reaches its maximum, the current also reaches its maximum (current and voltage are oscillating in phase). But for a capacitor or inductor, the maximum current flow occurs as the voltage passes through zero and vice versa (current and voltage are oscillating 90° out of phase,). Complex numbers are used to keep track of both the phase and magnitude of current and voltage:

$$V(t) = \text{Re}(V_0 e^{j\omega t}), \quad I(t) = \text{Re}(I_0 e^{j\omega t}), \quad Z = \frac{V_0}{I_0}, \quad Y = \frac{I_0}{V_0}$$

where:

- t is time,

- $V(t)$ and $I(t)$ are, respectively, voltage and current as a function of time,

- V_o, I_o, Z, and Y are complex numbers,

- Z is called impedance,

- Y is called admittance,

- Re indicates real part,

- ω is the angular frequency of the AC current,

- $j = \sqrt{-1}$ is the imaginary unit.

The impedance and admittance may be expressed as complex numbers that can be broken into real and imaginary parts:

$$Z = R + jX, \quad Y = G + jB$$

where R and G are resistance and conductance respectively, X is reactance, and B is susceptance. For ideal resistors, Z and Y reduce to R and G respectively, but for AC networks containing capacitors and inductors, X and B are nonzero.

$Z = 1/Y$ for AC circuits, just as R=1/G for DC circuits.

Frequency Dependence of Resistance

A key feature of AC circuits is that the resistance and conductance can be frequency-dependent, a phenomenon known as the universal dielectric response . One reason, mentioned above is the skin effect (and the related proximity effect). Another reason is that the resistivity itself may depend on frequency.

Energy Dissipation and Joule Heating

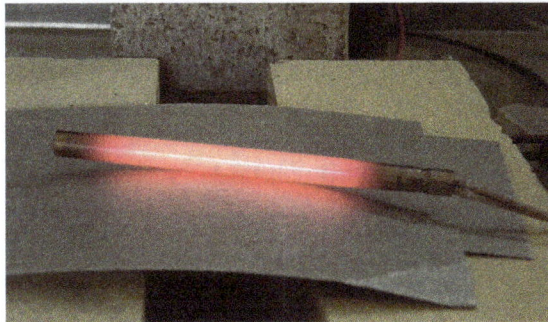

Running current through a material with high resistance creates heat, in a phenomenon called Joule heating. In this picture, a cartridge heater, warmed by Joule heating, is glowing red hot.

Resistors (and other elements with resistance) oppose the flow of electric current; therefore, electrical energy is required to push current through the resistance. This electrical energy is dissipated, heating the resistor in the process. This is called *Joule heating* (after James Prescott Joule), also called *ohmic heating* or *resistive heating*.

The dissipation of electrical energy is often undesired, particularly in the case of transmission losses in power lines. High voltage transmission helps reduce the losses by reducing the current for a given power.

On the other hand, Joule heating is sometimes useful, for example in electric stoves and other electric heaters (also called *resistive heaters*). As another example, incandescent lamps rely on Joule heating: the filament is heated to such a high temperature that it glows "white hot" with thermal radiation (also called incandescence).

The formula for Joule heating is:

$$P = I^2 R$$

where P is the power (energy per unit time) converted from electrical energy to thermal energy, R is the resistance, and I is the current through the resistor.

Dependence of Resistance on other Conditions

Temperature Dependence

Near room temperature, the resistivity of metals typically increases as temperature is increased, while the resistivity of semiconductors typically decreases as temperature is increased. The resistivity of insulators and electrolytes may increase or decrease depending on the system.

As a consequence, the resistance of wires, resistors, and other components often change with temperature. This effect may be undesired, causing an electronic circuit to malfunction at extreme temperatures. In some cases, however, the effect is put to good use. When temperature-dependent resistance of a component is used purposefully, the component is called a resistance thermometer or thermistor. (A resistance thermometer is made of metal, usually platinum, while a thermistor is made of ceramic or polymer.)

Resistance thermometers and thermistors are generally used in two ways: First, they can be used as thermometers: By measuring the resistance, the temperature of the environment can be inferred. Second, they can be used in conjunction with Joule heating (also called self-heating): If a large current is running through the resistor, the resistor's temperature rises and therefore its resistance changes. Therefore, these components can be used in a circuit-protection role similar to fuses, or for feedback in circuits, or for many other purposes. In general, self-heating can turn a resistor into a nonlinear and hysteretic circuit element.

If the temperature T does not vary too much, a linear approximation is typically used:

$$R(T) = R_0[1 + \alpha(T - T_0)]$$

where α is called the *temperature coefficient of resistance*, T_0 is a fixed reference temperature (usually room temperature), and R_0 is the resistance at temperature T_0. The parameter α is an empirical parameter fitted from measurement data. Because the linear approximation is only an approximation, α is different for different reference temperatures. For this reason it is usual to specify the temperature that α was measured at with a suffix, such as α_{15}, and the relationship only holds in a range of temperatures around the reference.

The temperature coefficient α is typically $+3 \times 10^{-3}$ K^{-1} to $+6 \times 10^{-3}$ K^{-1} for metals near room temperature. It is usually negative for semiconductors and insulators, with highly variable magnitude.

Strain Dependence

Just as the resistance of a conductor depends upon temperature, the resistance of a conductor depends upon strain. By placing a conductor under tension (a form of stress that leads to strain in the form of stretching of the conductor), the length of the section of conductor under tension increases and its cross-sectional area decreases. Both these effects contribute to increasing the resistance of the strained section of conductor. Under compression (strain in the opposite direction), the resistance of the strained section of conductor decreases.

Light Illumination Dependence

Some resistors, particularly those made from semiconductors, exhibit *photoconductivity*, meaning that their resistance changes when light is shining on them. Therefore, they are called *photoresistors* (or *light dependent resistors*). These are a common type of light detector.

Superconductivity

Superconductors are materials that have exactly zero resistance and infinite conductance, because they can have V=0 and I≠0. This also means there is no joule heating, or in other words no dissipation of electrical energy. Therefore, if superconductive wire is made into a closed loop, current flows around the loop forever. Superconductors require cooling to temperatures near 4 K with liquid helium for most metallic superconductors like niobium–tin alloys, or cooling to temperatures near 77K with liquid nitrogen for the expensive, brittle and delicate ceramic high temperature superconductors. Nevertheless, there are many technological applications of superconductivity, including superconducting magnets.

Electrical Resistivity and Conductivity

Electrical resistivity (also called specific electrical resistance or volume resistivity) and its inverse, electrical conductivity, is a fundamental property of a material that quantifies how strongly it resists or conducts electric current. A low resistivity indicates a material that readily allows electric current. Resistivity is commonly represented by the Greek letter ρ (rho). The SI unit of electrical resistivity is the ohm-meter ($\Omega \cdot m$). For example, if a 1 m × 1 m × 1 m solid cube of material has sheet contacts on two opposite faces, and the resistance between these contacts is 1 Ω, then the resistivity of the material is 1 $\Omega \cdot m$.

Electrical conductivity or specific conductance is the reciprocal of electrical resistivity. It represents a material's ability to conduct electric current. It is commonly signified by the Greek letter σ (sigma), but κ (kappa) (especially in electrical engineering) and γ (gamma) are sometimes used. The SI unit of electrical conductivity is siemens per metre (S/m).

Ideal Case

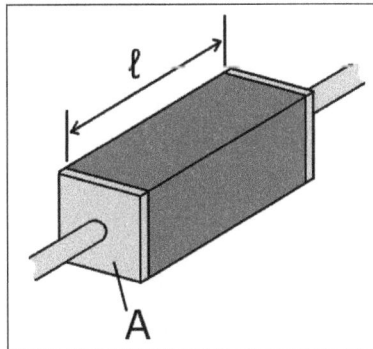

A piece of resistive material with electrical contacts on both ends.

In an ideal case, cross-section and physical composition of the examined material are uniform across the sample, and the electric field and current density are both parallel and constant everywhere. Many resistors and conductors do in fact have a uniform cross section with a uniform flow of electric current, and are made of a single material, so that this is a good model. When this is the case, the electrical resistivity ρ can be calculated by:

$$\rho = R\frac{A}{\ell},$$

where

      R is the electrical resistance of a uniform specimen of the material

      ℓ is the length of the specimen

A is the cross-sectional area of the specimen

Both *resistance* and *resistivity* describe how difficult it is to make electrical current flow through a material, but unlike resistance, resistivity is an *intrinsic property*. This means that all pure copper wires (which have not been subjected to distortion of their crystalline structure etc.), irrespective of their shape and size, have the same *resistivity*, but a long, thin copper wire has a much larger *resistance* than a thick, short copper wire. Every material has its own characteristic resistivity. For example, rubber has a far larger resistivity than copper.

In a hydraulic analogy, passing current through a high-resistivity material is like pushing water through a pipe full of sand—while passing current through a low-resistivity material is like pushing water through an empty pipe. If the pipes are the same size and shape, the pipe full of sand has higher resistance to flow. Resistance, however, is not *solely* determined by the presence or absence of sand. It also depends on the length and width of the pipe: short or wide pipes have lower resistance than narrow or long pipes.

The above equation can be transposed to get Pouillet's law (named after Claude Pouillet):

$$R = \rho\frac{\ell}{A}.$$

The resistance of a given material is proportional to the length, but inversely proportional to the cross-sectional area. Thus resistivity can be expressed using the SI unit "ohm metre" ($\Omega \cdot m$) — *i.e.* ohms divided by metres (for the length) and then multiplied by square metres (for the cross-sectional area).

For example, if $A = 1\ m^2$ $\ell = 1\ m$ (forming a cube with perfectly conductive contacts on opposite faces), then the resistance of this element in ohms is numerically equal to the resistivity of the material it is made of in $\Omega \cdot m$.

Conductivity, σ, is the inverse of resistivity:

$$\sigma = \frac{1}{\rho}.$$

Conductivity has SI units of "siemens per metre" (S/m).

General Scalar Case

For less ideal cases, such as more complicated geometry, or when the current and electric field vary in different parts of the material, it is necessary to use a more general expression in which the resistivity at a particular point is defined as the ratio of the electric field to the density of the current it creates at that point:

$$\rho = \frac{E}{J},$$

where,

ρ is the resistivity of the conductor material,

E is the magnitude of the electric field,

J is the magnitude of the current density,

in which E and J are inside the conductor.

Conductivity is the inverse (reciprocal) of resistivity. Here, it is given by:

$$\sigma = \frac{1}{\rho} = \frac{J}{E}.$$

For example, rubber is a material with large ρ and small σ—because even a very large electric field in rubber makes almost no current flow through it. On the other hand, copper is a material with small ρ and large σ—because even a small electric field pulls a lot of current through it.

As shown below, this expression simplifies to a single number when the electric field and current density are constant in the material.

Tensor Resistivity

When the resistivity of a material has a directional component, the most general definition of resistivity must be used. It starts from the tensor-vector form of Ohm's law, which relates the electric field inside a material to the electric current flow. This equation is completely general, meaning it is valid in all cases. However, this definition is the most complicated, so it is only directly used in anisotropic cases, where the more simple definitions cannot be applied. If the material is not anisotropic, it is safe to ignore the tensor-vector definition, and use a simpler expression instead.

Here, anisotropic means that the material has different properties in different directions. For example, a crystal of graphite consists microscopically of a stack of sheets, and current flows very easily through each sheet, but much less easily from one sheet to the adjacent one. In such cases, the current does not flow in exactly the same direction as the electric field. Thus, the appropriate equations are generalized to the three-dimensional tensor form:

$$J = \sigma E \rightleftharpoons E = \rho J$$

where the conductivity σ and resistivity ρ are rank-2 tensors, and electric field E and current density J are vectors. These tensors can be represented by 3×3 matrices, the vectors with 3x1 matrices, with matrix multiplication used on the right side of these equations. In matrix form, the resistivity relation is given by:

$$\begin{bmatrix} E_x \\ E_y \\ E_z \end{bmatrix} = \begin{bmatrix} \rho_{xx} & \rho_{xy} & \rho_{xz} \\ \rho_{yx} & \rho_{yy} & \rho_{yz} \\ \rho_{zx} & \rho_{zy} & \rho_{zz} \end{bmatrix} \begin{bmatrix} J_x \\ J_y \\ J_z \end{bmatrix}$$

where

E is the electric field vector, with components (E_x, E_y, E_z).

ρ is the resistivity tensor, in general a three by three matrix.

J is the electric current density vector, with components (J_x, J_y, J_z)

Equivalently, resistivity can be given in the more compact Einstein notation:

$$E_i = \rho_{ij} J_j$$

In either case, the resulting expression for each electric field component is:

$$E_x = \rho_{xx} J_x + \rho_{xy} J_y + \rho_{xz} J_z.$$
$$E_y = \rho_{yx} J_x + \rho_{yy} J_y + \rho_{yz} J_z.$$
$$E_z = \rho_{zx} J_x + \rho_{zy} J_y + \rho_{zz} J_z.$$

Since the choice of the coordinate system is free, the usual convention is to simplify the expression by choosing an x-axis parallel to the current direction, so $J_y = J_z = 0$. This leaves:

$$\rho_{xx} = \frac{E_x}{J_x}, \quad \rho_{yx} = \frac{E_y}{J_x}, \text{ and } \rho_{zx} = \frac{E_z}{J_x}.$$

Conductivity is defined similarly:

$$\begin{bmatrix} J_x \\ J_y \\ J_z \end{bmatrix} = \begin{bmatrix} \sigma_{xx} & \sigma_{xy} & \sigma_{xz} \\ \sigma_{yx} & \sigma_{yy} & \sigma_{yz} \\ \sigma_{zx} & \sigma_{zy} & \sigma_{zz} \end{bmatrix} \begin{bmatrix} E_x \\ E_y \\ E_z \end{bmatrix}$$

Or

$$J_i = \sigma_{ij} E_j$$

Both resulting in:

$$J_x = \sigma_{xx} E_x + \sigma_{xy} E_y + \sigma_{xz} E_z$$
$$J_y = \sigma_{yx} E_x + \sigma_{yy} E_y + \sigma_{yz} E_z$$
$$J_z = \sigma_{zx} E_x + \sigma_{zy} E_y + \sigma_{zz} E_z$$

Looking at the two expressions, ρ and σ are the matrix inverse of each other. However, in the most general case, the individual matrix elements are not necessarily reciprocals of one another; for example, σ_{xx} may not be equal to $1/\rho_{xx}$. This can be seen in the Hall effect, where ρ_{xy} is nonzero.

In the Hall effect, due to rotational invariance about the z-axis, $\rho_{yy} = \rho_{xx}$ and $\rho_{yx} = -\rho_{xy}$, so the relation between resistivity and conductivity simplifies to:

$$\sigma_{xx} = \frac{\rho_{xx}}{\rho_{xx}^2 + \rho_{xy}^2}, \quad \sigma_{xy} = \frac{-\rho_{xy}}{\rho_{xx}^2 + \rho_{xy}^2}$$

If the electric field is parallel to the applied current, ρ_{xy} and ρ_{xz} are zero. When they are zero, one number, ρ_{xx}, is enough to describe the electrical resistivity. It is then written as simply ρ, and this reduces to the simpler expression.

Causes of Conductivity

Band Theory Simplified

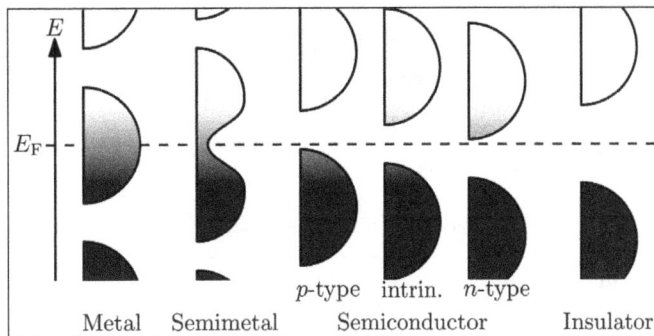

Filling of the electronic states in various types of materials at equilibrium: Here, height is energy while width is the density of available states for a certain energy in the material listed. The shade follows the Fermi–Dirac distribution (black = all states filled, white = no state filled). In metals and semimetals the Fermi level E_F lies inside at least one band. In insulators and semiconductors the Fermi level is inside a band gap; however, in semiconductors the bands are near enough to the Fermi level to be thermally populated with electrons or holes.

According to elementary quantum mechanics, an electron in an atom or crystal can only have certain precise energy levels; energies between these levels are impossible. When a large number of such allowed levels have close-spaced energy values – i.e. have energies that differ only minutely – those close energy levels in combination are called an "energy band". There can be many such energy bands in a material, depending on the atomic number of the constituent atoms and their distribution within the crystal.

The material's electrons seek to minimize the total energy in the material by settling into low energy states; however, the Pauli exclusion principle means that only one can exist in each such state. So the electrons "fill up" the band structure starting from the bottom. The characteristic energy level up to which the electrons have filled is called the Fermi level. The position of the Fermi level with respect to the band structure is very important for electrical conduction: Only electrons in energy levels near or above the Fermi level are free to move within the broader material structure, since the electrons can easily jump among the partially occupied states in that region. In contrast, the low energy states are completely filled with a fixed limit on the number of electrons at all times, and the high energy states are empty of electrons at all times.

Electric current consists of a flow of electrons. In metals there are many electron energy levels near the Fermi level, so there are many electrons available to move. This is what causes the high electronic conductivity of metals.

An important part of band theory is that there may be forbidden bands of energy: energy intervals that contain no energy levels. In insulators and semiconductors, the number of electrons is just the right amount to fill a certain integer number of low energy bands, exactly to the boundary. In this case, the Fermi level falls within a band gap. Since there are no available states near the Fermi level, and the electrons are not freely movable, the electronic conductivity is very low.

In Metals

Like balls in a Newton's cradle, electrons in a metal quickly transfer energy
from one terminal to another, despite their own negligible movement.

A metal consists of a lattice of atoms, each with an outer shell of electrons that freely dissociate from their parent atoms and travel through the lattice. This is also known as a positive ionic lattice. This 'sea' of dissociable electrons allows the metal to conduct electric current. When an electrical potential difference (a voltage) is applied across the metal, the resulting electric field causes electrons to drift towards the positive terminal. The actual drift velocity of electrons is typically small, on the order of magnitude of meters per hour. However, due to the sheer number of moving electrons, even a slow drift velocity results in a large current density. The mechanism is similar to transfer of momentum of balls in a Newton's cradle but the rapid propagation of an electric energy along a wire is not due to the mechanical forces, but the propagation of an energy-carrying electromagnetic field guided by the wire.

Most metals have electrical resistance. In simpler models (non quantum mechanical models) this can be explained by replacing electrons and the crystal lattice by a wave-like structure. When the electron wave travels through the lattice, the waves interfere, which causes resistance. The more regular the lattice is, the less disturbance happens and thus the less resistance. The amount of resistance is thus mainly caused by two factors. First, it is caused by the temperature and thus amount of vibration of the crystal lattice. Higher temperatures cause bigger vibrations, which act as irregularities in the lattice. Second, the purity of the metal is relevant as a mixture of different ions is also an irregularity.

In Semiconductors and Insulators

In metals, the Fermi level lies in the conduction band giving rise to free conduction electrons. However, in semiconductors the position of the Fermi level is within the band gap, about halfway between the conduction band minimum (the bottom of the first band of unfilled electron energy levels) and the valence band maximum (the top of the band below the conduction band, of filled electron energy levels). That applies for intrinsic (undoped) semiconductors. This means that at absolute zero temperature, there would be no free conduction electrons, and the resistance is infinite. However, the resistance decreases as the charge carrier density (i.e., without introducing further complications, the density of electrons) in the conduction band increases. In extrinsic (doped) semiconductors, dopant atoms increase the majority charge carrier concentration by donating electrons to the conduction band or producing holes in the valence band. (A "hole" is a position where an electron is missing; such holes can behave in a similar way to electrons.) For both types of donor or acceptor atoms, increasing dopant density reduces resistance. Hence, highly doped semiconductors behave metallically. At very high temperatures, the contribution of thermally generated carriers dominates over the contribution from dopant atoms, and the resistance decreases exponentially with temperature.

In Ionic Liquids/Electrolytes

In electrolytes, electrical conduction happens not by band electrons or holes, but by full atomic species (ions) traveling, each carrying an electrical charge. The resistivity of ionic solutions (electrolytes) varies tremendously with concentration – while distilled water is almost an insulator, salt water is a reasonable electrical conductor. Conduction in ionic liquids is also controlled by the movement of ions, but here we are talking about molten salts rather than solvated ions. In biological membranes, currents are carried by ionic salts. Small holes in cell membranes, called ion channels, are selective to specific ions and determine the membrane resistance.

Superconductivity

The electrical resistivity of a metallic conductor decreases gradually as temperature is lowered. In ordinary conductors, such as copper or silver, this decrease is limited by impurities and other defects. Even near absolute zero, a real sample of a normal conductor shows some resistance. In a superconductor, the resistance drops abruptly to zero when the material is cooled below its critical temperature. An electric current flowing in a loop of superconducting wire can persist indefinitely with no power source.

In 1986, researchers discovered that some cuprate-perovskite ceramic materials have much higher critical temperatures, and in 1987 one was produced with a critical temperature above 90 K (−183 °C). Such a high transition temperature is theoretically impossible for a conventional superconductor, so the researchers named these conductors *high-temperature superconductors*. Liquid nitrogen boils at 77 K, cold enough to activate high-temperature superconductors, but not nearly cold enough for conventional superconductors. In conventional superconductors, electrons are held together in pairs by an attraction mediated by lattice phonons. The best available model of high-temperature superconductivity is still somewhat crude. There is a hypothesis that electron pairing in high-temperature superconductors is mediated by short-range spin waves known as paramagnons.

Plasma

Plasmas are very good conductors and electric potentials play an important role. The potential as it exists on average in the space between charged particles, independent of the question of how it can be measured, is called the *plasma potential*, or *space potential*. If an electrode is inserted into a plasma, its potential generally lies considerably below the plasma potential, due to what is termed a Debye sheath. The good electrical conductivity of plasmas makes their electric fields very small. This results in the important concept of *quasineutrality*, which says the density of negative charges is approximately equal to the density of positive charges over large volumes of the plasma (n_e = <Z>n_i), but on the scale of the Debye length there can be charge imbalance. In the special case that *double layers* are formed, the charge separation can extend some tens of Debye lengths.

Lightning is an example of plasma present at Earth's surface. Typically, lightning discharges 30,000 amperes at up to 100 million volts, and emits light, radio waves, and X-rays. Plasma temperatures in lightning might approach 30,000 Kelvin (29,727 °C) (53,540 °F), or five times hotter than the temperature at the sun surface, and electron densities may exceed 10^{24} m^{-3}.

The magnitude of the potentials and electric fields must be determined by means other than simply finding the net charge density. A common example is to assume that the electrons satisfy the Boltzmann relation:

$$n_e \propto e^{e\Phi/k_B T_e}.$$

Differentiating this relation provides a means to calculate the electric field from the density:

$$E = -\frac{k_B T_e}{e} \frac{\nabla n_e}{n_e}.$$

(The "downward pointing triangle" is a vector gradient; see nabla symbol and gradient for more information.)

It is possible to produce a plasma that is not quasineutral. An electron beam, for example, has only negative charges. The density of a non-neutral plasma must generally be very low, or it must be very small. Otherwise, the repulsive electrostatic force dissipates it.

In astrophysical plasmas, Debye screening prevents electric fields from directly affecting the plasma over large distances, i.e., greater than the Debye length. However, the existence of charged particles causes the plasma to generate, and be affected by, magnetic fields. This can and does cause extremely complex behavior, such as the generation of plasma double layers, an object that separates charge over a few tens of Debye lengths. The dynamics of plasmas interacting with external and self-generated magnetic fields are studied in the academic discipline of magnetohydrodynamics.

Plasma is often called the *fourth state of matter* after solid, liquids and gases. It is distinct from these and other lower-energy states of matter. Although it is closely related to the gas phase in that it also has no definite form or volume, it differs in a number of ways, including the following:

Property	Gas	Plasma
Electrical conductivity	Very low: air is an excellent insulator until it breaks down into plasma at electric field strengths above 30 kilovolts per centimeter.	Usually very high: for many purposes, the conductivity of a plasma may be treated as infinite.
Independently acting species	One: all gas particles behave in a similar way, influenced by gravity and by collisions with one another.	Two or three: electrons, ions, protons and neutrons can be distinguished by the sign and value of their charge so that they behave independently in many circumstances, with different bulk velocities and temperatures, allowing phenomena such as new types of waves and instabilities.
Velocity distribution	Maxwellian: collisions usually lead to a Maxwellian velocity distribution of all gas particles, with very few relatively fast particles.	Often non-Maxwellian: collisional interactions are often weak in hot plasmas and external forcing can drive the plasma far from local equilibrium and lead to a significant population of unusually fast particles.
Interactions	Binary: two-particle collisions are the rule, three-body collisions extremely rare.	Collective: waves, or organized motion of plasma, are very important because the particles can interact at long ranges through the electric and magnetic forces.

Resistivity and Conductivity of Various Materials

- A conductor such as a metal has high conductivity and a low resistivity.

- An insulator like glass has low conductivity and a high resistivity.

- The conductivity of a semiconductor is generally intermediate, but varies widely under different conditions, such as exposure of the material to electric fields or specific frequencies of light, and, most important, with temperature and composition of the semiconductor material.

The degree of doping in semiconductors makes a large difference in conductivity. To a point, more doping leads to higher conductivity. The conductivity of a solution of water is highly dependent on its concentration of dissolved salts, and other chemical species that ionize in the solution. Electrical conductivity of water samples is used as an indicator of how salt-free, ion-free, or impurity-free the sample is; the purer the water, the lower the conductivity (the higher the resistivity). Conductivity measurements in water are often reported as *specific conductance*, relative to the conductivity of pure water at 25 °C. An EC meter is normally used to measure conductivity in a solution.

Material	Resistivity, ρ ($\Omega \cdot$m)
Superconductors	0
Metals	10^{-8}
Semiconductors	Variable
Electrolytes	Variable
Insulators	10^{16}
Superinsulators	∞

The effective temperature coefficient varies with temperature and purity level of the material. The 20 °C value is only an approximation when used at other temperatures. For example, the coefficient becomes lower at higher temperatures for copper, and the value 0.00427 is commonly specified at 0 °C.

The extremely low resistivity (high conductivity) of silver is characteristic of metals. George Gamow tidily summed up the nature of the metals' dealings with electrons in his popular science book *One, Two, Three...Infinity* (1947):

> The metallic substances differ from all other materials by the fact that the outer shells of their atoms are bound rather loosely, and often let one of their electrons go free. Thus the interior of a metal is filled up with a large number of unattached electrons that travel aimlessly around like a crowd of displaced persons. When a metal wire is subjected to electric force applied on its opposite ends, these free electrons rush in the direction of the force, thus forming what we call an electric current.

More technically, the free electron model gives a basic description of electron flow in metals.

Wood is widely regarded as an extremely good insulator, but its resistivity is sensitively dependent on moisture content, with damp wood being a factor of at least 10^{10} worse insulator than oven-dry. In any case, a sufficiently high voltage – such as that in lightning strikes or some high-tension power lines – can lead to insulation breakdown and electrocution risk even with apparently dry wood.

Temperature Dependence

Linear Approximation

The electrical resistivity of most materials changes with temperature. If the temperature T does not vary too much, a linear approximation is typically used:

$$\rho(T) = \rho_0[1 + \alpha(T - T_0)]$$

where α is called the *temperature coefficient of resistivity*, T_0 is a fixed reference temperature (usually room temperature), and ρ_0 is the resistivity at temperature T_0. The parameter α is an empirical parameter fitted from measurement data. Because the linear approximation is only an approximation, α is different for different reference temperatures. For this reason it is usual to specify the temperature that α was measured at with a suffix, such as α_{15}, and the relationship only holds in a range of temperatures around the reference. When the temperature varies over a large temperature range, the linear approximation is inadequate and a more detailed analysis and understanding should be used.

Metals

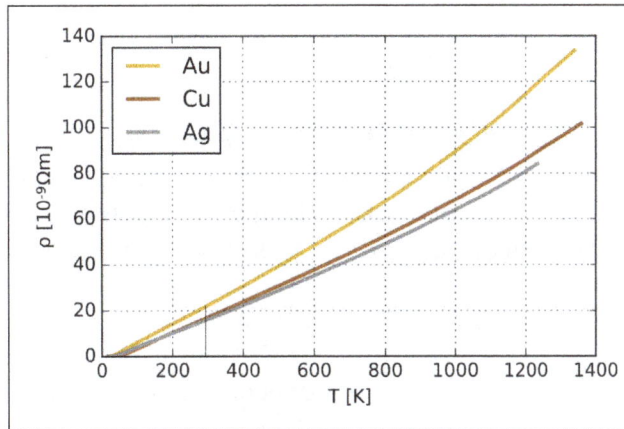

Temperature dependence of the resistivity of gold, copper and silver.

In general, electrical resistivity of metals increases with temperature. Electron–phonon interactions can play a key role. At high temperatures, the resistance of a metal increases linearly with temperature. As the temperature of a metal is reduced, the temperature dependence of resistivity follows a power law function of temperature. Mathematically the temperature dependence of the resistivity ρ of a metal is given by the Bloch–Grüneisen formula:

$$\rho(T) = \rho(0) + A\left(\frac{T}{\Theta_R}\right)^n \int_0^{\Theta_R/T} \frac{x^n}{(e^x-1)(1-e^{-x})}\,dx$$

where $\rho(0)$ is the residual resistivity due to defect scattering, A is a constant that depends on the velocity of electrons at the Fermi surface, the Debye radius and the number density of electrons in the metal. Θ_R is the Debye temperature as obtained from resistivity measurements and matches very closely with the values of Debye temperature obtained from specific heat measurements. n is an integer that depends upon the nature of interaction:

- $n = 5$ implies that the resistance is due to scattering of electrons by phonons (as it is for simple metals).

- $n = 3$ implies that the resistance is due to s-d electron scattering (as is the case for transition metals).

- $n = 2$ implies that the resistance is due to electron–electron interaction.

If more than one source of scattering is simultaneously present, Matthiessen's Rule (first formulated by Augustus Matthiessen in the 1860s) states that the total resistance can be approximated by adding up several different terms, each with the appropriate value of n.

As the temperature of the metal is sufficiently reduced (so as to 'freeze' all the phonons), the resistivity usually reaches a constant value, known as the residual resistivity. This value depends not only on the type of metal, but on its purity and thermal history. The value of the residual resistivity of a metal is decided by its impurity concentration. Some materials lose all electrical resistivity at sufficiently low temperatures, due to an effect known as superconductivity.

An investigation of the low-temperature resistivity of metals was the motivation to Heike Kamerlingh Onnes's experiments that led in 1911 to discovery of superconductivity.

Semiconductors

In general, intrinsic semiconductor resistivity decreases with increasing temperature. The electrons are bumped to the conduction energy band by thermal energy, where they flow freely, and in doing so leave behind holes in the valence band, which also flow freely. The electric resistance of a typical intrinsic (non doped) semiconductor decreases exponentially with temperature:

$$\rho = \rho_0 e^{-aT}$$

An even better approximation of the temperature dependence of the resistivity of a semiconductor is given by the Steinhart–Hart equation:

$$\frac{1}{T} = A + B \ln \rho + C(\ln \rho)^3$$

where A, B and C are the so-called Steinhart–Hart coefficients.

This equation is used to calibrate thermistors.

Extrinsic (doped) semiconductors have a far more complicated temperature profile. As temperature increases starting from absolute zero they first decrease steeply in resistance as the carriers leave the donors or acceptors. After most of the donors or acceptors have lost their carriers, the resistance starts to increase again slightly due to the reducing mobility of carriers (much as in a metal). At higher temperatures, they behave like intrinsic semiconductors as the carriers from the donors/acceptors become insignificant compared to the thermally generated carriers.

In non-crystalline semiconductors, conduction can occur by charges quantum tunnelling from one localised site to another. This is known as variable range hopping and has the characteristic form of,

$$\rho = A \exp\left(T^{-1/n}\right),$$

where $n = 2, 3, 4$, depending on the dimensionality of the system.

Complex Resistivity and Conductivity

When analyzing the response of materials to alternating electric fields (dielectric spectroscopy), in applications such as electrical impedance tomography, it is convenient to replace resistivity with a complex quantity called impedivity (in analogy to electrical impedance). Impedivity is the sum of a real component, the resistivity, and an imaginary component, the reactivity (in analogy to reactance). The magnitude of impedivity is the square root of sum of squares of magnitudes of resistivity and reactivity.

Conversely, in such cases the conductivity must be expressed as a complex number (or even as a matrix of complex numbers, in the case of anisotropic materials) called the *admittivity*. Admittivity is the sum of a real component called the conductivity and an imaginary component called the susceptivity.

An alternative description of the response to alternating currents uses a real (but frequency-dependent) conductivity, along with a real permittivity. The larger the conductivity is, the more quickly the alternating-current signal is absorbed by the material (i.e., the more opaque the material is).

Kirchhoff's Circuit Laws

Kirchhoff's circuit laws are two equalities that deal with the current and potential difference (commonly known as voltage) in the lumped element model of electrical circuits. They were first described in 1845 by German physicist Gustav Kirchhoff. This generalized the work of Georg Ohm and preceded the work of James Clerk Maxwell. Widely used in electrical engineering, they are also called Kirchhoff's rules or simply Kirchhoff's laws. These laws can be applied in time and frequency domains and form the basis for network analysis.

Both of Kirchhoff's laws can be understood as corollaries of Maxwell's equations in the low-frequency limit. They are accurate for DC circuits, and for AC circuits at frequencies where the wavelengths of electromagnetic radiation are very large compared to the circuits.

Kirchhoff's Current Law

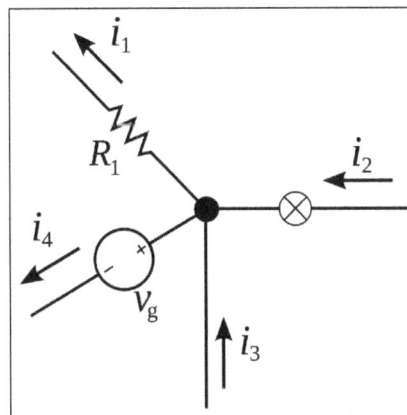

The current entering any junction is equal to the current leaving that junction. $i_2 + i_3 = i_1 + i_4$.

This law is also called Kirchhoff's first law, Kirchhoff's point rule, or Kirchhoff's junction rule (or nodal rule).

This law states that, for any node (junction) in an electrical circuit, the sum of currents flowing into that node is equal to the sum of currents flowing out of that node; or equivalently:

The algebraic sum of currents in a network of conductors meeting at a point is zero.

Recalling that current is a signed (positive or negative) quantity reflecting direction towards or away from a node, this principle can be succinctly stated as:

$$\sum_{k=1}^{n} I_k = 0$$

where n is the total number of branches with currents flowing towards or away from the node.

The law is based on the conservation of charge where the charge (measured in coulombs) is the product of the current (in amperes) and the time (in seconds). If the net charge in a region is constant, the current law will hold on the boundaries of the region. This means that the current law relies on the fact that the net charge in the wires and components is constant..

Uses

A matrix version of Kirchhoff's current law is the basis of most circuit simulation software, such as SPICE. The current law is used with Ohm's law to perform nodal analysis.

The current law is applicable to any lumped network irrespective of the nature of the network; whether unilateral or bilateral, active or passive, linear or non-linear.

Kirchhoff's Voltage Law

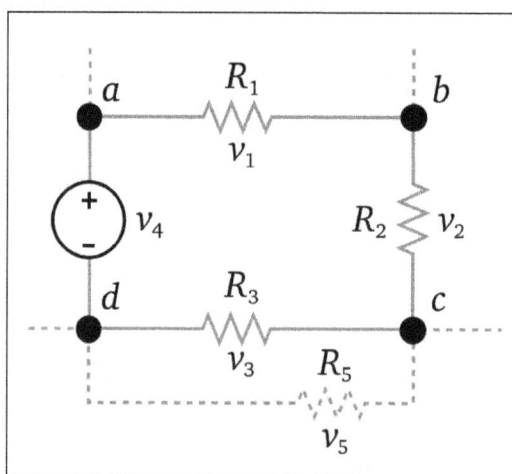

The sum of all the voltages around a loop is equal to zero.
$$v_1 + v_2 + v_3 + v_4 = 0.$$

This law is also called Kirchhoff's second law, Kirchhoff's loop (or mesh) rule, and Kirchhoff's second rule.

This law states that:

> The Directed Sum of the Potential Differences (Voltages) Around any Closed Loop is Zero.

Similarly to Kirchhoff's current law, the voltage law can be stated as:

$$\sum_{k=1}^{n} V_k = 0$$

Here, n is the total number of voltages measured.

Derivation of Kirchhoff's Voltage Law

Consider some arbitrary circuit. Approximate the circuit with lumped elements, so that (time-varying) magnetic fields are contained to each component and the field in the region exterior to the circuit is negligible. Based on this assumption, the Maxwell-Faraday equation reveals that,

$$\nabla \times E = -\frac{\partial B}{\partial t} = 0$$

in the exterior region. If each of the components have a finite volume, then the exterior region is simply connected, and thus the electric field is conservative in that region. Therefore, for any loop in the circuit, we find that,

$$\sum V_i = -\sum \int_{\mathcal{P}i} E \cdot dl = \oint E \cdot dl = 0$$

where $\mathcal{P}_i$ are paths around the *exterior* of each of the components, from one terminal to another.

Generalization

In the low-frequency limit, the voltage drop around any loop is zero. This includes imaginary loops arranged arbitrarily in space – not limited to the loops delineated by the circuit elements and conductors. In the low-frequency limit, this is a corollary of Faraday's law of induction (which is one of Maxwell's equations).

This has practical application in situations involving "static electricity".

Limitations

Kirchhoff's circuit laws are the result of the lumped element model and both depend on the model being applicable to the circuit in question. When the model is not applicable, the laws do not apply.

The current law is dependent on the assumption that the net charge in any wire, junction or lumped component is constant. Whenever the electric field between parts of the circuit is non-negligible, such as when two wires are capacitively coupled, this may not be the case. This occurs in high-frequency AC circuits, where the lumped element model is no longer applicable. For example, in a transmission line, the charge density in the conductor will constantly be oscillating.

On the other hand, the voltage law relies on the fact that the action of time-varying magnetic fields are confined to individual components, such as inductors. In reality, the induced electric field produced by an inductor is not confined, but the leaked fields are often negligible.

Modelling Real Circuits with Lumped Elements

The lumped element approximation for a circuit is accurate at low frequencies. At higher frequencies, leaked fluxes and varying charge densities in conductors become significant. To an extent, it is possible to still model such circuits using parasitic components. If frequencies are too high, it may be more appropriate to simulate the fields directly using finite element modelling or other techniques.

To model circuits so that both laws can still be used, it is important to understand the distinction between *physical* circuit elements and the *ideal* lumped elements. For example, a wire is not an ideal conductor. Unlike an ideal conductor, wires can inductively and capacitively couple to each other (and to themselves), and have a finite propagation delay. Real conductors can be modeled in terms of lumped elements by considering parasitic capacitances distributed between the conductors to model capacitive coupling, or parasitic (mutual) inductances to model inductive coupling. Wires also have some self-inductance, which is the reason that decoupling capacitors are necessary.

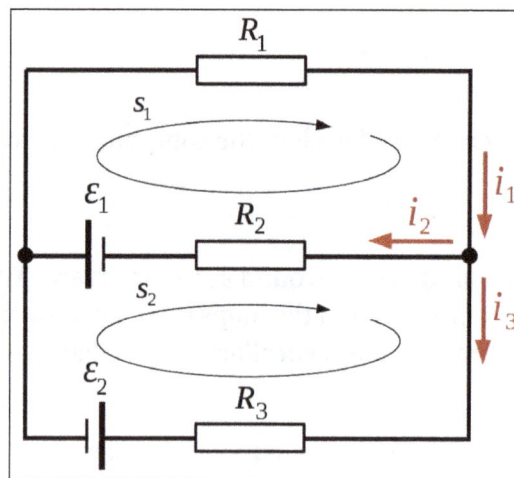

Assume an electric network consisting of two voltage sources and three resistors.

According to the first law:

$$i_1 - i_2 - i_3 = 0$$

Applying the second law to the closed circuit s_1, and substituting for voltage using Ohm's law gives:

$$-R_2 i_2 + \mathcal{E}_1 - R_1 i_1 = 0$$

The second law, again combined with Ohm's law, applied to the closed circuit s_2 gives:

$$-R_3 i_3 - \mathcal{E}_2 - \mathcal{E}_1 + R_2 i_2 = 0$$

This yields a system of linear equations in i_1, i_2, i_3 :

$$\begin{cases} i_1 - i_2 - i_3 & = 0 \\ -R_2 i_2 + \mathcal{E}_1 - R_1 i_1 & = 0 \\ -R_3 i_3 - \mathcal{E}_2 - \mathcal{E}_1 + R_2 i_2 & = 0 \end{cases}$$

which is equivalent to,

$$\begin{cases} i_1 + (-i_2) + (-i_3) & = 0 \\ R_1 i_1 + R_2 i_2 + 0 i_3 & = \mathcal{E}_1 \\ 0 i_1 + R_2 i_2 - R_3 i_3 & = \mathcal{E}_1 + \mathcal{E}_2 \end{cases}$$

Assuming,

$$R_1 = 100\Omega,\ R_2 = 200\Omega,\ R_3 = 300\Omega$$
$$\mathcal{E}_1 = 3V, \mathcal{E}_2 = 4V$$

the solution is,

$$\begin{cases} i_1 = \dfrac{1}{1100} A \\ i_2 = \dfrac{4}{275} A \\ i_3 = -\dfrac{3}{220} A \end{cases}$$

The current i_3 has a negative sign which means the assumed direction of i_3 was incorrect and i_3 is actually flowing in the direction opposite to the red arrow labeled i_3. The current in R_3 flows from left to right.

Thévenin's Theorem

As originally stated in terms of DC resistive circuits only, Thévenin's theorem (aka Helmholtz–Thévenin theorem) holds that:

- Any linear electrical network containing only voltage sources, current sources and resistances can be replaced at terminals A-B by an equivalent combination of a voltage source V_{th} in a series connection with a resistance R_{th}.

- The equivalent voltage V_{th} is the voltage obtained at terminals A-B of the network with terminals A-B open circuited.

- The equivalent resistance R_{th} is the resistance that the circuit between terminals A and B would have if all ideal voltage sources in the circuit were replaced by a short circuit and all ideal current sources were replaced by an open circuit.

- If terminals A and B are connected to one another, the current flowing from A to B will be V_{th}/R_{th}. This means that R_{th} could alternatively be calculated as V_{th} divided by the short-circuit current between A and B when they are connected together.

In circuit theory terms, the theorem allows any one-port network to be reduced to a single voltage source and a single impedance.

The theorem also applies to frequency domain AC circuits consisting of reactive and resistive impedances. It means the theorem applies for AC in an exactly same way to DC except that resistances are generalized to impedances.

The theorem was independently derived in 1853 by the German scientist Hermann von Helmholtz and in 1883 by Léon Charles Thévenin (1857–1926), an electrical engineer with France's national Postes et Télégraphes telecommunications organization.

Thévenin's theorem and its dual, Norton's theorem, are widely used to make circuit analysis simpler and to study a circuit's initial-condition and steady-state response. Thévenin's theorem can be used to convert any circuit's sources and impedances to a Thévenin equivalent; use of the theorem may in some cases be more convenient than use of Kirchhoff's circuit laws.

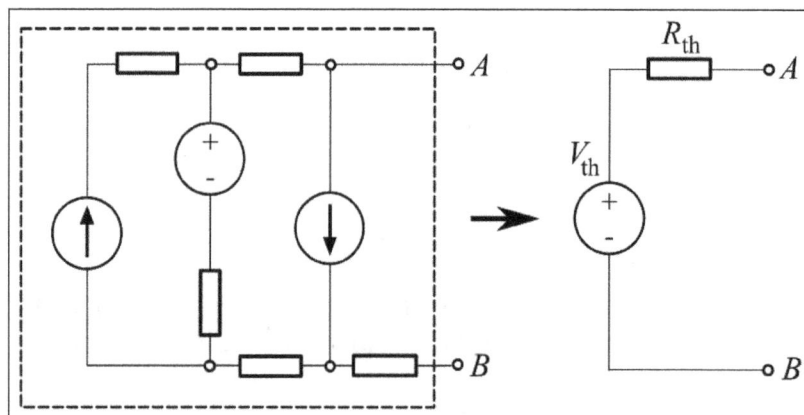

Any black box containing resistances only and voltage and current sources can be replaced by a Thévenin equivalent circuit consisting of an equivalent voltage source in series connection with an equivalent resistance.

Calculating the Thévenin Equivalent

The equivalent circuit is a voltage source with voltage V_{Th} in series with a resistance R_{Th}.

The Thévenin-equivalent voltage V_{Th} is the open-circuit voltage at the output terminals of the original circuit. When calculating a Thévenin-equivalent voltage, the voltage divider principle is often useful, by declaring one terminal to be V_{out} and the other terminal to be at the ground point.

The Thévenin-equivalent resistance R_{Th} is the resistance measured across points A and B "looking back" into the circuit. The resistance is measured after replacing all voltage- and current-sources with their internal resistances. That means an ideal voltage source is replaced with a short circuit, and an ideal current source is replaced with an open circuit. Resistance can then be calculated across the terminals using the formulae for series and parallel circuits. This method is valid only for circuits with independent sources. If there are dependent sources in the circuit, another method must be used such as connecting a test source across A and B and calculating the voltage across or current through the test source.

The replacements of voltage and current sources do what the sources would do if their values were set to zero. A zero valued voltage source would create a potential difference of zero volts between its terminals, regardless of the current that passes through it; its replacement, a short circuit, does the same thing. A zero valued current source passes zero current, regardless of the voltage across it; its replacement, an open circuit, does the same thing.

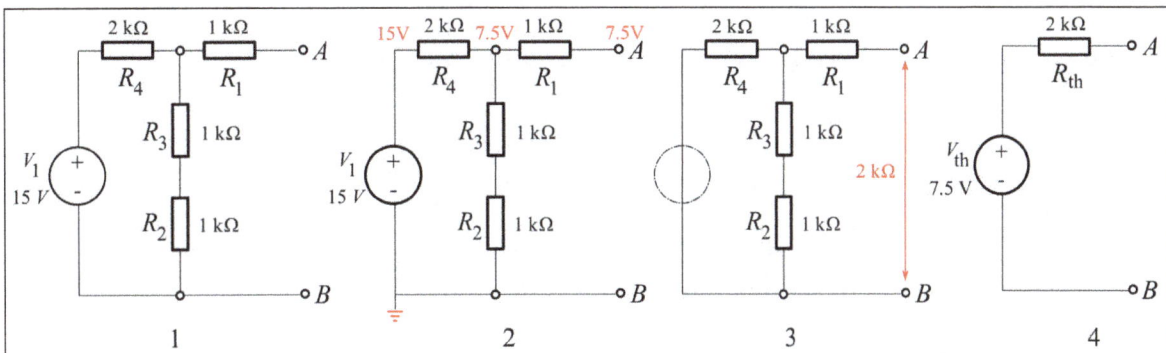

1. Original circuit, 2. The equivalent voltage,
3. The equivalent resistance, 4. The equivalent circuit

In the example, calculating the equivalent voltage:

$$V_{Th} = \frac{R_2 + R_3}{(R_2 + R_3) + R_4} \cdot V_1$$

$$= \frac{1k\Omega + 1k\Omega}{(1k\Omega + 1k\Omega) + 2k\Omega} \cdot 15\,V$$

$$= \frac{1}{2} \cdot 15\,V = 7.5\,V$$

(notice that R_1 is not taken into consideration, as above calculations are done in an open-circuit condition between A and B, therefore no current flows through this part, which means there is no current through R_1 and therefore no voltage drop along this part). Calculating equivalent resistance ($R_x \setminus R_y$ is the total resistance of two parallel resistors):

$$R_{Th} = R_1 + \left[(R_2 + R_3) \| R_4\right]$$

$$= 1k\Omega + \left[(1k\Omega + 1k\Omega) \| 2k\Omega\right]$$

$$= 1k\Omega + \left(\frac{1}{(1k\Omega + 1k\Omega)} + \frac{1}{(2k\Omega)}\right)^{-1} = 2k\Omega.$$

Conversion to a Norton Equivalent

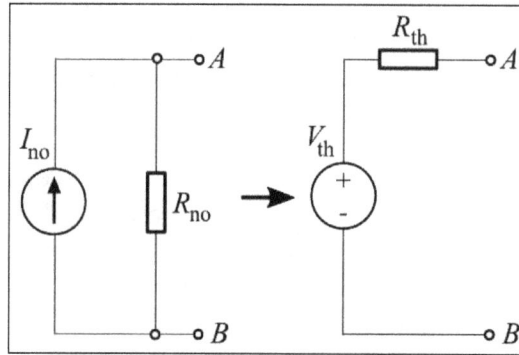

Norton-Thevenin conversion

A Norton equivalent circuit is related to the Thévenin equivalent by,

$$R_{Th} = R_{No}$$
$$V_{Th} = I_{No} R_{No}$$
$$I_{No} = V_{Th} / R_{Th}$$

Practical Limitations

- Many circuits are only linear over a certain range of values, thus the Thévenin equivalent is valid only within this linear range.

- The Thévenin equivalent has an equivalent I–V characteristic only from the point of view of the load.

- The power dissipation of the Thévenin equivalent is not necessarily identical to the power dissipation of the real system. However, the power dissipated by an external resistor between the two output terminals is the same regardless of how the internal circuit is implemented.

A Proof of the Theorem

The proof involves two steps. The first step is to use superposition theorem to construct a solution. Then, uniqueness theorem is employed to show that the obtained solution is unique. It is noted that the second step is usually implied in literature.

By using superposition of specific configurations, it can be shown that for any linear "black box" circuit which contains voltage sources and resistors, its voltage is a linear function of the corresponding current as follows:

$$V = V_{Eq} - Z_{Eq} I.$$

Here, the first term reflects the linear summation of contributions from each voltage source, while the second term measures the contributions from all the resistors. The above expression is obtained by using the fact that the voltage of the black box for a given current I is identical to the linear superposition of the solutions of the following problems: (1) to leave the black box open circuited but activate individual voltage source one at a time and, (2) to short circuit all the voltage sources but feed the circuit with a certain ideal voltage source so that the resulting current exactly reads I (Alternatively, one can use an ideal current source of current I). Moreover, it is straightforward to show that V_{Eq} and Z_{Eq} are the single voltage source and the single series resistor in question.

As a matter of fact, the above relation between V and I is established by superposition of some particular configurations. Now, the uniqueness theorem guarantees that the result is general. To be specific, there is one and only one value of V once the value of I is given. In other words, the above relation holds true independent of what the "black box" is plugged to.

Norton's Theorem

Norton's theorem (aka Mayer–Norton theorem) holds, to illustrate in DC circuit theory terms:

- Any linear electrical network with voltage and current sources and only resistances can be replaced at terminals A–B by an equivalent current source I_{no} in parallel connection with an equivalent resistance R_{no}.

- This equivalent current I_{no} is the current obtained at terminals A-B of the network with terminals A-B short circuited.

- This equivalent resistance R_{no} is the resistance obtained at terminals A-B of the network with all its voltage sources short circuited and all its current sources open circuited.

For alternating current (AC) systems the theorem can be applied to reactive impedances as well as resistances.

The Norton equivalent circuit is used to represent any network of linear sources and impedances at a given frequency.

Norton's theorem and its dual, Thévenin's theorem, are widely used for circuit analysis simplification and to study circuit's initial-condition and steady-state response.

Norton's theorem was independently derived in 1926 by Siemens & Halske researcher Hans Ferdinand Mayer and Bell Labs engineer Edward Lawry Norton.

To find the equivalent:

- Find the Norton current I_{no}. Calculate the output current, I_{AB}, with a short circuit as the load (meaning 0 resistance between A and B). This is I_{no}.

- Find the Norton resistance R_{no}. When there are no dependent sources (all current and voltage sources are independent), there are two methods of determining the Norton impedance R_{no}.

- Calculate the output voltage, V_{AB}, when in open circuit condition (i.e., no load resistor – meaning infinite load resistance). R_{no} equals this V_{AB} divided by I_{no}.

or

- • Replace independent voltage sources with short circuits and independent current sources with open circuits. The total resistance across the output port is the Norton impedance R_{no}.

This is equivalent to calculating the Thevenin resistance. However, when there are dependent sources, the more general method must be used. This method is not shown below in the diagrams.

- • Connect a constant current source at the output terminals of the circuit with a value of 1 ampere and calculate the voltage at its terminals. This voltage divided by the 1 A current is the Norton impedance R_{no}. This method must be used if the circuit contains dependent sources, but it can be used in all cases even when there are no dependent sources.

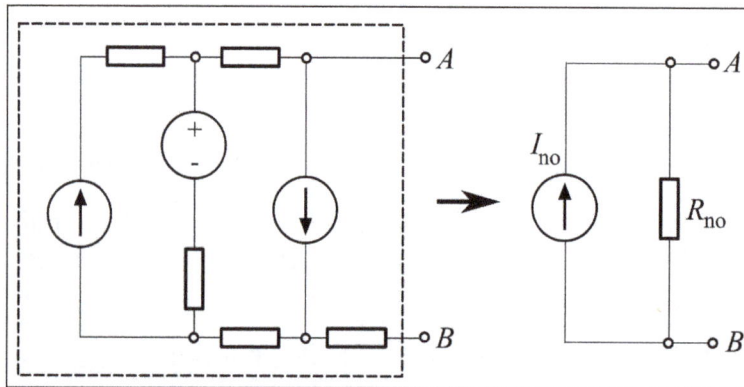

Any black box containing resistances only and voltage and current sources can be replaced by an equivalent circuit consisting of an equivalent current source in parallel connection with an equivalent resistance.

Example of a Norton Equivalent Circuit

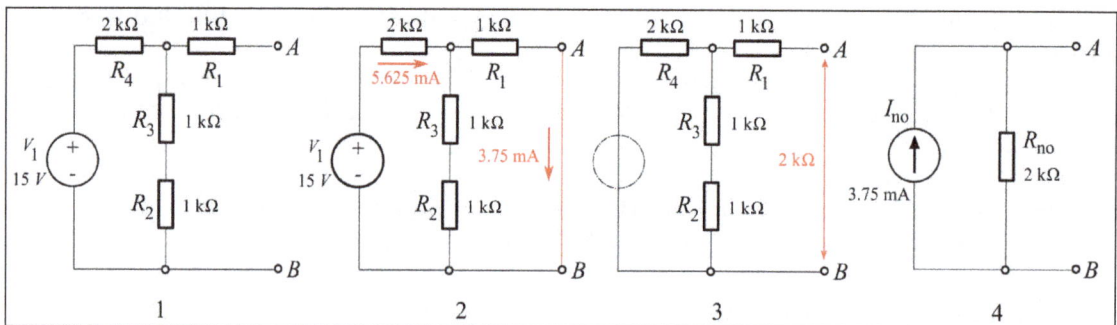

Figure: 1. The original circuit, 2. Calculating the equivalent output current,
3. Calculating the equivalent resistance, 4. Design the Norton equivalent circuit

In the example, the total current I_{total} is given by:

$$I_{total} = \frac{15\,V}{2\,k\Omega + 1\,k\Omega \parallel (1\,k\Omega + 1\,k\Omega)} = 5.625\,mA.$$

The current through the load is then, using the current divider rule:

$$I_{no} = \frac{1\,k\Omega + 1\,k\Omega}{(1\,k\Omega + 1\,k\Omega + 1\,k\Omega)} \cdot I_{total}$$

$$= 2/3 \cdot 5.625\,\text{mA} = 3.75\,\text{mA}.$$

And the equivalent resistance looking back into the circuit is:

$$R_{no} = 1\,\text{k}\Omega + (2\,\text{k}\Omega \parallel (1\,\text{k}\Omega + 1\,\text{k}\Omega)) = 2\,\text{k}\Omega.$$

So the equivalent circuit is a 3.75 mA current source in parallel with a 2 kΩ resistor.

Conversion to a Thévenin Equivalent

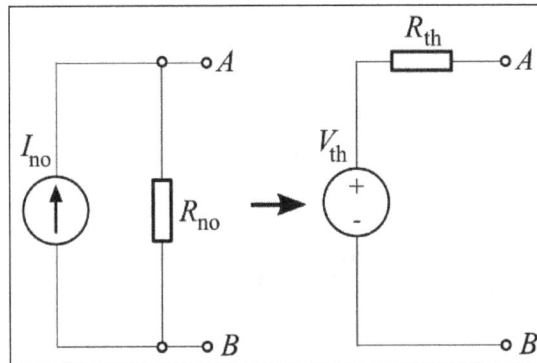

To a Thévenin equivalent.

A Norton equivalent circuit is related to the Thévenin equivalent by the equations:

$$R_{th} = R_{no}$$
$$V_{th} = I_{no} R_{no}$$
$$\frac{V_{th}}{R_{th}} = I_{no}$$

Superposition Theorem

The superposition theorem for electrical circuits states that for a linear system the response (voltage or current) in any branch of a bilateral linear circuit having more than one independent source equals the algebraic sum of the responses caused by each independent source acting alone, where all the other independent sources are replaced by their internal impedances.

To ascertain the contribution of each individual source, all of the other sources first must be "turned off" (set to zero) by:

- Replacing all other independent voltage sources with a short circuit (thereby eliminating difference of potential i.e. *V*=0; internal impedance of ideal voltage source is zero (short circuit)).

- Replacing all other independent current sources with an open circuit (thereby eliminating current i.e. *I*=0; internal impedance of ideal current source is infinite (open circuit)).

This procedure is followed for each source in turn, then the resultant responses are added to determine the true operation of the circuit. The resultant circuit operation is the superposition of the various voltage and current sources.

The superposition theorem is very important in circuit analysis. It is used in converting any circuit into its Norton equivalent or Thevenin equivalent.

The theorem is applicable to linear networks (time varying or time invariant) consisting of independent sources, linear dependent sources, linear passive elements (resistors, inductors, capacitors) and linear transformers.

Superposition works for voltage and current but not power. In other words, the sum of the powers of each source with the other sources turned off is not the real consumed power. To calculate power we first use superposition to find both current and voltage of each linear element and then calculate the sum of the multiplied voltages and currents.

Maximum Power Transfer Theorem

In electrical engineering, the maximum power transfer theorem states that, to obtain *maximum* external power from a source with a finite internal resistance, the resistance of the load must equal the resistance of the source as viewed from its output terminals. Moritz von Jacobi published the maximum power (transfer) theorem around 1840; it is also referred to as "Jacobi's law".

The theorem results in maximum *power* transfer across the circuit, and not maximum *efficiency*. If the resistance of the load is made larger than the resistance of the source, then efficiency is higher, since a higher percentage of the source power is transferred to the load, but the *magnitude* of the load power is lower since the total circuit resistance goes up.

If the load resistance is smaller than the source resistance, then most of the power ends up being dissipated in the source, and although the total power dissipated is higher, due to a lower total resistance, it turns out that the amount dissipated in the load is reduced.

The theorem states how to choose (so as to maximize power transfer) the load resistance, once the source resistance is given. It is a common misconception to apply the theorem in the opposite scenario. It does *not* say how to choose the source resistance for a given load resistance. In fact, the source resistance that maximizes power transfer is always zero, regardless of the value of the load resistance.

The theorem can be extended to alternating current circuits that include reactance, and states that maximum power transfer occurs when the load impedance is equal to the complex conjugate of the source impedance.

Maximizing Power Transfer versus Power Efficiency

The theorem was originally misunderstood (notably by Joule) to imply that a system consisting of an electric motor driven by a battery could not be more than 50% efficient since, when the impedances were matched, the power lost as heat in the battery would always be equal to the power er delivered to the motor.

In 1880, this assumption was shown to be false by either Edison or his colleague Francis Robbins Upton, who realized that maximum efficiency was not the same as maximum power transfer.

To achieve maximum efficiency, the resistance of the source (whether a battery or a dynamo) could be (or should be) made as close to zero as possible. Using this new understanding, they obtained an efficiency of about 90%, and proved that the electric motor was a practical alternative to the heat engine.

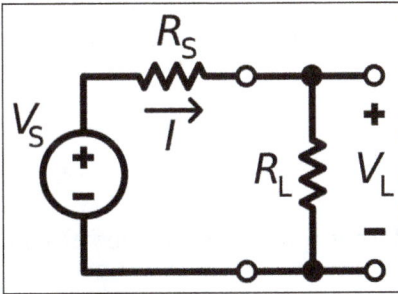

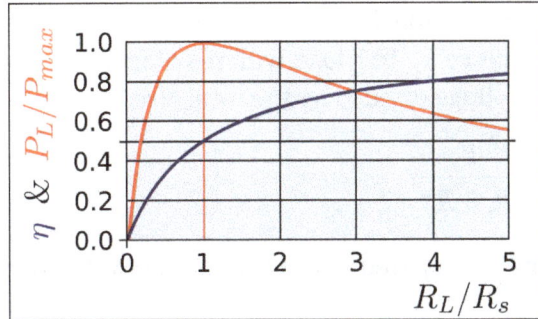

The condition of maximum power transfer does not result in maximum efficiency.

If we define the efficiency η as the ratio of power dissipated by the load, R_L, to power developed by the source, V_s, then it is straightforward to calculate from the above circuit diagram that:

$$\eta = \frac{R_L}{R_L + R_S} = \frac{1}{1 + R_S / R_L}.$$

Consider three particular cases:

- If $R_L = R_S$, then $\eta = 0.5$,

- If $R_L \to \infty$ or $R_S = 0$, then $\eta = 1$,

- If $R_L = 0$, then $\eta = 0$.

The efficiency is only 50% when maximum power transfer is achieved, but approaches 100% as the load resistance approaches infinity, though the total power level tends towards zero.

Efficiency also approaches 100% if the source resistance approaches zero, and 0% if the load resistance approaches zero. In the latter case, all the power is consumed inside the source (unless the source also has no resistance), so the power dissipated in a short circuit is zero.

Impedance Matching

A related concept is reflectionless impedance matching. In radio frequency transmission lines, and other electronics, there is often a requirement to match the source impedance (at the transmitter) to the load impedance (such as an antenna) to avoid reflections in the transmission line that could overload or damage the transmitter.

Calculus-based Proof for Purely Resistive Circuits

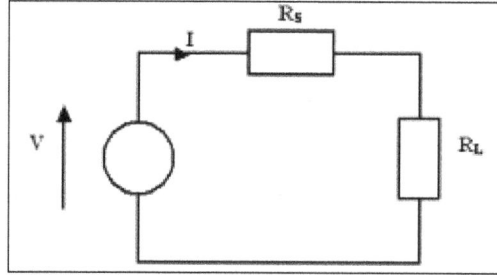

In the diagram opposite, power is being transferred from the source, with voltage V and fixed source resistance R_S, to a load with resistance R_L, resulting in a current I. By Ohm's law, I is simply the source voltage divided by the total circuit resistance:

$$I = \frac{V}{R_S + R_L}.$$

The power P_L dissipated in the load is the square of the current multiplied by the resistance:

$$P_L = I\ R_L = \left(\frac{V}{R_S + R_L}\right) R_L = \frac{V}{R_S / R_L + 2R_S + R_L}$$

The value of R_L for which this expression is a maximum could be calculated by differentiating it, but it is easier to calculate the value of R_L for which the denominator,

$$R_S^2 / R_L + 2R_S + R_L$$

is a minimum. The result will be the same in either case. Differentiating the denominator with respect to R_L:

$$\frac{d}{dR_L}\left(R_S^2 / R_L + 2R_S + R_L\right) = -R_S^2 / R_L^2 + 1.$$

For a maximum or minimum, the first derivative is zero, so

$$R_S^2 / R_L^2 = 1$$

Or

$$R_L = \pm R_S.$$

In practical resistive circuits, R_S and R_L are both positive, so the positive sign in the above is the correct solution.

To find out whether this solution is a minimum or a maximum, the denominator expression is differentiated again:

$$\frac{d^2}{dR_L^2}\left(R_S^2 / R_L + 2R_S + R_L\right) = 2R_S^2 / R_L^3.$$

This is always positive for positive values of R_S and R_L, showing that the denominator is a minimum, and the power is therefore a maximum, when

$R_S = R_L$.

The above proof assumes fixed source resistance R_S. When the source resistance can be varied, power transferred to the load can be increased by reducing R_S. For example, a 100 Volt source with an R_S of 10Ω will deliver 250 watts of power to a 10Ω load; reducing R_S to 0Ω increases the power delivered to 1000 watts.

Note that this shows that maximum power transfer can also be interpreted as the load voltage being equal to one-half of the Thevenin voltage equivalent of the source.

In Reactive Circuits

The power transfer theorem also applies when the source and/or load are not purely resistive.

A refinement of the maximum power theorem says that any reactive components of source and load should be of equal magnitude but opposite sign.

- This means that the source and load impedances should be *complex conjugates* of each other.

- In the case of purely resistive circuits, the two concepts are identical.

Physically realizable sources and loads are not usually purely resistive, having some inductive or capacitive components, and so practical applications of this theorem, under the name of complex conjugate impedance matching, do, in fact, exist.

If the source is totally inductive (capacitive), then a totally capacitive (inductive) load, in the absence of resistive losses, would receive 100% of the energy from the source but send it back after a quarter cycle.

The resultant circuit is nothing other than a resonant LC circuit in which the energy continues to oscillate to and from. This oscillation is called reactive power.

Power factor correction (where an inductive reactance is used to "balance out" a capacitive one), is essentially the same idea as complex conjugate impedance matching although it is done for entirely different reasons.

For a fixed reactive *source*, the maximum power theorem maximizes the real power (P) delivered to the load by complex conjugate matching the load to the source.

For a fixed reactive *load*, power factor correction minimizes the apparent power (S) (and unnecessary current) conducted by the transmission lines, while maintaining the same amount of real power transfer.

This is done by adding a reactance to the load to balance out the load's own reactance, changing the reactive load impedance into a resistive load impedance.

Proof:

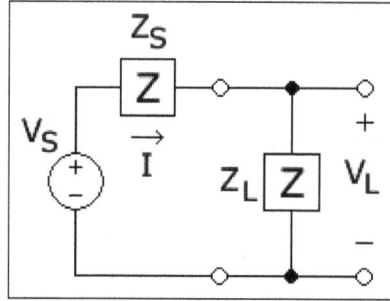

In this diagram, AC power is being transferred from the source, with phasor magnitude of voltage $|V_S|$ (positive peak voltage) and fixed source impedance Z_S (S for source), to a load with impedance Z_L (L for load), resulting in a (positive) magnitude $|I|$ of the current phasor I. This magnitude $|I|$ results from dividing the magnitude of the source voltage by the magnitude of the total circuit impedance:

$$|I| = \frac{|V_S|}{|Z_S + Z_L|}.$$

The average power P_L dissipated in the load is the square of the current multiplied by the resistive portion (the real part) R_L of the load impedance Z_L:

$$P_L = I_{rms}^2 R_L = \frac{1}{2}|I|^2 R_L$$

$$= \frac{1}{2}\left(\frac{|V_S|}{|Z_S + Z_L|}\right)^2 R_L = \frac{1}{2}\frac{|V_S|^2 R_L}{(R_S + R_L)^2 + (X_S + X_L)^2},$$

where R_S and R_L denote the resistances, that is the real parts, and X_S and X_L denote the reactances, that is the imaginary parts, of respectively the source and load impedances Z_S and Z_L.

To determine, for a given source voltage V_S and impedance Z_S, the value of the load impedance Z_L for which this expression for the power yields a maximum, one first finds, for each fixed positive value of R_L, the value of the reactive term X_L for which the denominator,

$$(R_S + R_L)^2 + (X_S + X_L)^2$$

is a minimum. Since reactances can be negative, this is achieved by adapting the load reactance to

$$X_L = -X_S.$$

This reduces the above equation to:

$$P_L = \frac{1}{2}\frac{|V_S|^2 R_L}{(R_S + R_L)^2}$$

and it remains to find the value of R_L which maximizes this expression. This problem has the same form as in the purely resistive case, and the maximizing condition therefore is $R_L = R_S$.

The two maximizing conditions:

- $R_L = R_S$
- $X_L = -X_S$

describe the complex conjugate of the source impedance, denoted by *, and thus can be concisely combined to:

$$Z_L = Z_S^*.$$

DC Circuit

The closed path in which the direct current flows is called the DC circuit. The current flows in only one direction and it is mostly used in low voltage applications. The resistor is the main component of the DC circuit.

A simple DC circuit is shown in the figure below which contains a DC source (battery), a load lamp, a switch, connecting leads, and measuring instruments like ammeter and voltmeter. The load resistor is connected in series, parallel or series-parallel combination as per requirement.

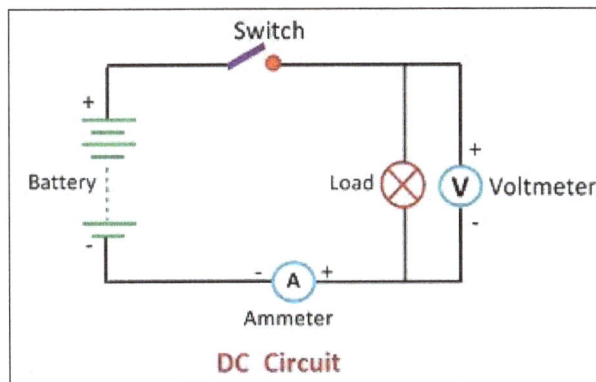

Types of DC Circuit

The DC electric circuit is mainly classified into three groups: They are the series DC circuit, parallel DC circuit, and series and parallel DC circuit.

DC Series Circuit

The circuit in which have DC series source, and the number of resistors are connected end to end so that same current flow through them is called a DC series circuit. The figure below shows the simple series circuit. In the series circuit the resistor R_1, R_2, and R_3 are connected in series across a supply voltage of V volts. The same current I is flowing through all the three resistors.

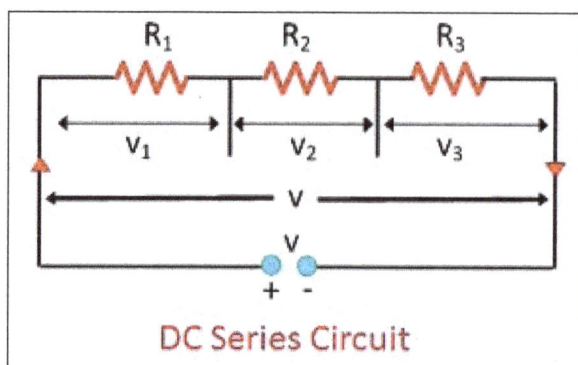

DC Series Circuit

If V_1, V_2, and V_3 are the voltage drop across the three resistor R_1, R_2, and R_3 respectively, then

$$V = V_1 + V_2 + V_3$$
$$V = IR_1 + IR_2 + IR_3$$

Let R be the total resistance of the circuit then,

$$IR = IR_1 + IR_2 + IR_3$$
$$R = R_1 + R_2 + R_3$$

Total resistance = Sum of the individual resistance.

In such type of circuit all the lamps are controlled by the single switch and they cannot be controlled individually. The most common application of this circuit is for decoration purpose where a number of low voltage lamps are connected in series.

DC Parallel Circuit

The circuit which have DC source and one end of all the resistors is joined to a common point and other end are also joined to another common point so that current flows through them is called a DC parallel circuit.

The figure shows a simple parallel circuit. In this circuit the three resistor R_1, R_2, and R_3 are connected in parallel across a supply voltage of V volts. The current flowing through them is I_1, I_2 and I_3 respectively.

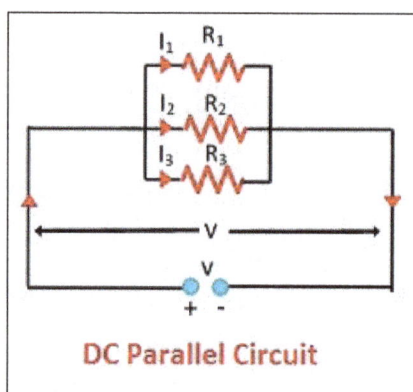

DC Parallel Circuit

The total current drawn by the circuit,

$$I = I_1 + I_2 + I_3$$

$$I = \frac{V}{R_1} + \frac{V}{R_2} + \frac{V}{R_3}$$

Let R be the total or effective resistance of the circuit, then,

$$\frac{V}{R} = \frac{V}{R_1} + \frac{V}{R_2} + \frac{V}{R_3}$$

$$\frac{1}{R} = \frac{1}{R_1} + \frac{1}{R_2} + \frac{1}{R_3}$$

Reciprocal of total resistance = sum of reciprocal of the individual resistance..

All the resistance is operated to the same voltage, therefore all of them are connected in parallel. Each of them can be controlled individually with the help of a separate switch.

DC Series-Parallel Circuit

The circuit in which series and parallel circuit are connected in series is called a series parallel circuit. The figure below show the series-parallel circuit. In this circuit, two resistor R_1 and R_2 are connected in parallel with each other across terminal AB. The other three resistors R_3, R_4 and R_6 are connected in parallel with each other across terminal BC.

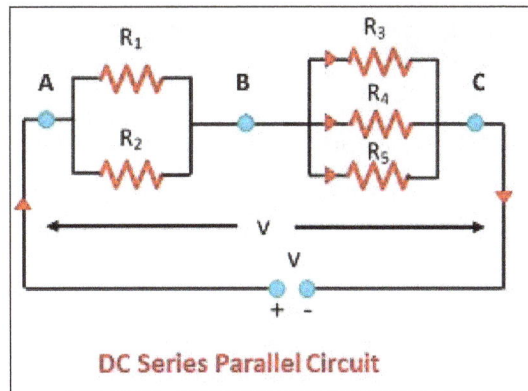

DC Series Parallel Circuit

The two groups of resistor R_{AB} and R_{BC} are connected in series with each other across the supply voltage of V volts. The total or effective resistance of the whole circuit can be determined as given below:

$$\frac{1}{R_{AB}} = \frac{1}{R_1} + \frac{1}{R_2} = \frac{R_1 + R_2}{R_1 R_2}$$

$$R_{AB} = \frac{R_1 R_2}{R_1 + R_2}$$

Similarly,

$$\frac{1}{R_{BC}} = \frac{1}{R_3} + \frac{1}{R_4} + \frac{1}{R_5} = \frac{R_3 R_4 + R_4 R_5 + R_5 R_3}{R_3 R_4 R_5}$$

$$R_{BC} = \frac{R_3 R_4 R_5}{R_3 R_4 + R_4 R_5 + R_5 R_3}$$

Total or effective resistance of the ciruit,

$$R = R_{AB} + R_{BC}$$

AC Circuit

In alternating current (AC) circuits, instead of a constant voltage supplied by a battery, the voltage oscillates in a sine wave pattern, varying with time as:

$$V = V_0 \sin \omega t$$

In a household circuit, the frequency is 60 Hz. The angular frequency is related to the frequency, f, by:

$$\omega = 2\pi f$$

Vo represents the maximum voltage, which in a household circuit in North America is about 170 volts. We talk of a household voltage of 120 volts, though; this number is a kind of average value of the voltage. The particular averaging method used is something called root mean square (square the voltage to make everything positive, find the average, take the square root), or rms. Voltages and currents for AC circuits are generally expressed as rms values. For a sine wave, the relationship between the peak and the rms average is:

rms value = 0.707 peak value

Resistance in an AC Circuit

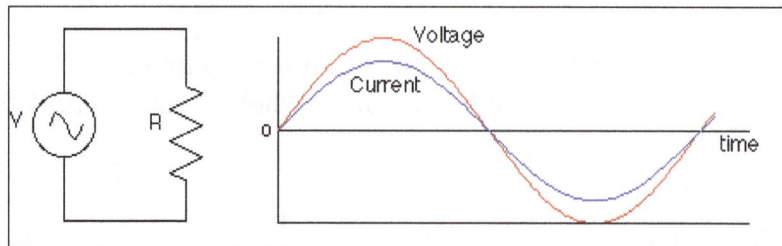

The relationship V = IR applies for resistors in an AC circuit, so

$$I = V / R = (V_0 / R)\sin(\omega t) = I_0 \sin(\omega t)$$

In AC circuits we'll talk a lot about the phase of the current relative to the voltage. In a circuit which only involves resistors, the current and voltage are in phase with each other, which means that the peak voltage is reached at the same instant as peak current. In circuits which have capacitors and inductors (coils) the phase relationships will be quite different.

Capacitance in an AC Circuit

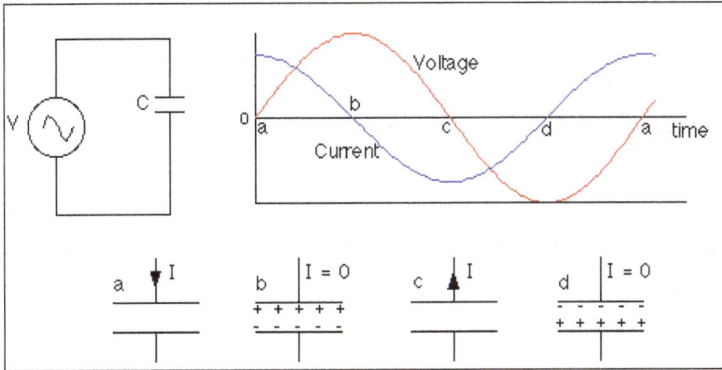

Consider now a circuit which has only a capacitor and an AC power source (such as a wall outlet). A capacitor is a device for storing charging. It turns out that there is a 90° phase difference between the current and voltage, with the current reaching its peak 90° (1/4 cycle) before the voltage reaches its peak. Put another way, the current leads the voltage by 90° in a purely capacitive circuit.

To understand why this is, we should review some of the relevant equations, including:

- Relationship between voltage and charge for a capacitor: $CV = Q$

- Relationship between current and the flow of charge: $I = \Delta Q / \Delta t$

The AC power supply produces an oscillating voltage. We should follow the circuit through one cycle of the voltage to figure out what happens to the current.

Step 1: At point a the voltage is zero and the capacitor is uncharged. Initially, the voltage increases quickly. The voltage across the capacitor matches the power supply voltage, so the current is large to build up charge on the capacitor plates. The closer the voltage gets to its peak, the slower it changes, meaning less current has to flow. When the voltage reaches a peak at point b, the capacitor is fully charged and the current is momentarily zero.

Step 2: After reaching a peak, the voltage starts dropping. The capacitor must discharge now, so the current reverses direction. When the voltage passes through zero at point c, it's changing quite rapidly; to match this voltage the current must be large and negative.

Step 3: Between points c and d, the voltage is negative. Charge builds up again on the capacitor plates, but the polarity is opposite to what it was in step one. Again the current is negative, and as the voltage reaches its negative peak at point d the current drops to zero.

Step 4: After point d, the voltage heads toward zero and the capacitor must discharge. When the voltage reaches zero it's gone through a full cycle so it's back to point a again to repeat the cycle.

The larger the capacitance of the capacitor, the more charge has to flow to build up a particular voltage on the plates, and the higher the current will be. The higher the frequency of the voltage, the shorter the time available to change the voltage, so the larger the current has to be. The current, then, increases as the capacitance increases and as the frequency increases.

Usually this is thought of in terms of the effective resistance of the capacitor, which is known as the capacitive reactance, measured in ohms. There is an inverse relationship between current and resistance, so the capacitive reactance is inversely proportional to the capacitance and the frequency:

A capacitor in an AC circuit exhibits a kind of resistance called capacitive reactance, measured in ohms. This depends on the frequency of the AC voltage, and is given by:

Capacitive reactance : $X_c = 1/\omega C = 1/2\pi f C$

We can use this like a resistance (because, really, it is a resistance) in an equation of the form V = IR to get the voltage across the capacitor:

$$V = I \ Xc$$

Note that V and I are generally the rms values of the voltage and current.

Inductance in an AC Circuit

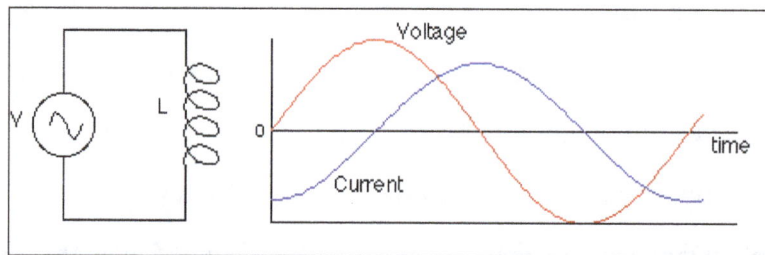

An inductor is simply a coil of wire (often wrapped around a piece of ferromagnet). If we now look at a circuit composed only of an inductor and an AC power source, we will again find that there is a 90° phase difference between the voltage and the current in the inductor. This time, however, the current lags the voltage by 90°, so it reaches its peak 1/4 cycle after the voltage peaks.

The reason for this has to do with the law of induction:

$$\varepsilon = -N\Delta\Phi/\Delta t \quad or \quad \varepsilon = -L\Delta I/\Delta t$$

Applying Kirchoff's loop rule to the circuit above gives:

$$V = L\Delta I/\Delta t = 0 \quad so \quad V = L \ \Delta I/\Delta t$$

As the voltage from the power source increases from zero, the voltage on the inductor matches it. With the capacitor, the voltage came from the charge stored on the capacitor plates (or, equivalently, from the electric field between the plates). With the inductor, the voltage comes from changing the flux through the coil, or, equivalently, changing the current through the coil, which changes the magnetic field in the coil.

To produce a large positive voltage, a large increase in current is required. When the voltage passes through zero, the current should stop changing just for an instant. When the voltage is large and negative, the current should be decreasing quickly. These conditions can all be satisfied by having the current vary like a negative cosine wave, when the voltage follows a sine wave.

How does the current through the inductor depend on the frequency and the inductance? If the frequency is raised, there is less time to change the voltage. If the time interval is reduced, the change in current is also reduced, so the current is lower. The current is also reduced if the inductance is increased.

As with the capacitor, this is usually put in terms of the effective resistance of the inductor. This effective resistance is known as the inductive reactance. This is given by:

$$X_L = \omega L = 2\pi f \, L$$

where L is the inductance of the coil (this depends on the geometry of the coil and whether its got a ferromagnetic core). The unit of inductance is the henry.

As with capacitive reactance, the voltage across the inductor is given by:

$$V = IX_L$$

References

- Srednicki, Mark A. (2007). Quantum field theory. Cambridge, [England] ; New York [NY.]: Cambridge University Press. ISBN 978-0-521-86449-7

- Electrodynamics, electrical, engineering, science, encyclopedia: infoplease.com, Retrieved 14 July, 2019

- Serway, Raymond A.; Jewett, John W., Jr. (2004). Physics for scientists and engineers, with modern physics. Belmont, [CA.]: Thomson Brooks/Cole. ISBN 0-534-40846-X

- Joseph Henry". Distinguished Members Gallery, National Academy of Sciences. Archived from the original on 2013-12-13. Retrieved 2006-11-30

- Raymond A. Serway, John W. Jewett (2006). Principles of Physics. Thomson Brooks/Cole. P. 807. ISBN 978-0-534-49143-7

- Liu, Changli (2017). "Explanation on Overdetermination of Maxwell's Equations". Physics and Engineering. 27 (3): 7–9. Arxiv:1002.0892. Bibcode:2010arxiv1002.0892L. Doi:10.3969/j.issn.1009-7104.2017.03.002

- Spencer, James N.; et al. (2010). Chemistry: Structure and Dynamics. John Wiley & Sons. P. 78. ISBN 9780470587119

- Electromagnetic-radiation,science: britannica.com, Retrieved 11 January, 2019

Permissions

All chapters in this book are published with permission under the Creative Commons Attribution Share Alike License or equivalent. Every chapter published in this book has been scrutinized by our experts. Their significance has been extensively debated. The topics covered herein carry significant information for a comprehensive understanding. They may even be implemented as practical applications or may be referred to as a beginning point for further studies.

We would like to thank the editorial team for lending their expertise to make the book truly unique. They have played a crucial role in the development of this book. Without their invaluable contributions this book wouldn't have been possible. They have made vital efforts to compile up to date information on the varied aspects of this subject to make this book a valuable addition to the collection of many professionals and students.

This book was conceptualized with the vision of imparting up-to-date and integrated information in this field. To ensure the same, a matchless editorial board was set up. Every individual on the board went through rigorous rounds of assessment to prove their worth. After which they invested a large part of their time researching and compiling the most relevant data for our readers.

The editorial board has been involved in producing this book since its inception. They have spent rigorous hours researching and exploring the diverse topics which have resulted in the successful publishing of this book. They have passed on their knowledge of decades through this book. To expedite this challenging task, the publisher supported the team at every step. A small team of assistant editors was also appointed to further simplify the editing procedure and attain best results for the readers.

Apart from the editorial board, the designing team has also invested a significant amount of their time in understanding the subject and creating the most relevant covers. They scrutinized every image to scout for the most suitable representation of the subject and create an appropriate cover for the book.

The publishing team has been an ardent support to the editorial, designing and production team. Their endless efforts to recruit the best for this project, has resulted in the accomplishment of this book. They are a veteran in the field of academics and their pool of knowledge is as vast as their experience in printing. Their expertise and guidance has proved useful at every step. Their uncompromising quality standards have made this book an exceptional effort. Their encouragement from time to time has been an inspiration for everyone.

The publisher and the editorial board hope that this book will prove to be a valuable piece of knowledge for students, practitioners and scholars across the globe.

Index